T0136995

Springer Atmospheric Sciences

More information about this series at http://www.springer.com/series/10176

Sivaramakrishnan Lakshmivarahan
John M. Lewis • Rafal Jabrzemski

Forecast Error Correction using Dynamic Data Assimilation

 Springer

Sivaramakrishnan Lakshmivarahan
School of Computer Science
University of Oklahoma
Norman, OK, USA

Rafal Jabrzemski
Oklahoma Climatological Survey
University of Oklahoma
Norman, OK, USA

John M. Lewis
National Severe Storms Laboratory
Norman, OK, USA

Desert Research Institute
Reno, NV, USA

ISSN 2194-5217   ISSN 2194-5225   (electronic)
Springer Atmospheric Sciences
ISBN 978-3-319-82010-1   ISBN 978-3-319-39997-3   (eBook)
DOI 10.1007/978-3-319-39997-3

Printed on acid-free paper

This Springer imprint is published by Springer Nature
The registered company is Springer International Publishing AG Switzerland

Yoshikazu Sasaki (1927–2015), Father of Variational Data Assimilation in Meteorology

Preface

In this book, we focus on deterministic forecasting based on governing dynamical equations (typically in the form of differential equations). These equations require specification of a control vector for their solution (initial conditions, boundary conditions, physical and/or empirical parameters). Defining forecast error is central to our study as the book title implies. An all-inclusive definition of forecast error is difficult to formulate. Therefore, we find it best to define this error categorically: (1) error due to incorrectly specified terms in the governing equations (or the absence of important terms in these equations), (2) inexact numerical approximations to the analytic form of the dynamical equations including artificial amplification/damping of solutions in the numerical integration process, and (3) uncertainty in the elements of control. That is, error in prediction results from incorrect dynamical laws, numerical inexactitude, and uncertainty in control. And the true test of forecast goodness rests on comparing the forecast with accurate observations. Ofttimes, we are unable to definitively determine the source(s) of error. But given this error, we ask the question: Can we improve the forecast by altering the control vector or empirically correcting the dynamical law? And just as important a question, can we determine the relative impact of the various elements of control on the forecast of interest?

Immediately we see that there is a desire to determine sensitivity of forecast to elements of control—labeled forecast uncertainty. This uncertainty is then used to find optimal elements of control—values of the elements that minimize the sum of squared differences between the forecast and observations and ideal placement of observations in space-time. Clearly, this path blends sensitivity with least squares fit of model to observations. A methodology that considers all of the factors mentioned above is called the forecast sensitivity method (FSM), a relatively new form of dynamic data assimilation that is the centerpiece of this book.

There is a rich history of work in both sensitivity analysis (SA) and dynamic data assimilation (DDA). Sensitivity analysis has been a mainstay of both dynamical systems and biostatistical systems. The works of Gregor Mendel and Ronald Fisher are excellent examples of sensitivity in the field of population genetics—necessarily in the form of statistics for hybridization of peas in Mendel's case and statistics of crop production in Fisher's case. Henri Poincaré was the pioneer in the uncertainty of dynamical forecasts with respect to elements of control. Even in the absence of computational power in the late nineteenth century, he clearly understood the extreme sensitivity of the three-body problem's solution to slight changes in initial conditions (presented as an exercise in this book). One of the most frequently quoted sentences in his studies was "La prediction deviant impossible" [The prediction becomes impossible]. Lorenz (1995) took this to mean that Poincaré was close to discovering the theory of chaos. These issues in deterministic forecasting have led to the investigation of predictability limits, that point in time when the forecast is no better than "climatology" (the average state of affairs in the system). In engineering and control theory, Hendrik Bode, a research mathematician at Bell Laboratories, developed sensitivity analysis in service to feedback control that he studied during WWII when he addressed problems of gunnery control. His classic work is titled *Network Analysis and Feedback Amplifier Design* (Bode et al. 1945). The two-volume treatise on *Sensitivity and Uncertainty Analysis* by Cacuci (2003) and Cacuci et al. (2005) deals extensively with the discussion of adjoint sensitivity and its applications to geosciences. For other applications of sensitivity analysis to engineering systems, refer to Cruz (1982), Deif (2012), Eslami (2013), Fiacco (1984), Frank (1978), Kokotovic and Rutman (1965), Ronen (1988), Rozenwasser and Yusupov (1999), Saltelli et al. (2000, 2004, 2008), and Sobral (1968).

In regard to DDA, celebrated mathematician Carl Gauss did his fundamental work on least squares fitting of observations to models in the first decade of the nineteenth century (Gauss 1809) (a discussion of his problem is found in this book). This method has never fallen out of use. Gauss's work was expanded from particle dynamics to continuous media through the work of Alfred Clebsch (Clebsch 1857). In the 1950s, Japanese meteorologist Yoshi Sasaki used Gauss' fundamental idea in combination with the Clebsch transformation to develop DDA for numerical weather prediction (NWP). This methodology has come to be called variational analysis in meteorology and in abbreviated form 4D-VAR (Sasaki 1958). Lewis et al. (2006) review variational analysis in the context of NWP. Also refer to Lewis and Lakshmivarahan (2008) for more details. A resource for numerous papers in sensitivity analysis can be found in Rabitz et al. (1983), Nago (1971) and Tomović and Vukobratovíc (1972).

As might be expected, there is equivalence between classic Gaussian least squares methodology (variational analysis) and FSM. We carefully examine this connection in the book and offer several problems (and demonstrations) that explore this connection along with advantages/disadvantages of these DDA methods.

The book is partitioned into two sections: Part I, a general theory of FSM with a variety of practical problems that give substance to the theory, and Part II, an in-depth analysis of FSM applied to well-known geophysical dynamics problems—the dynamics of shallow water waves and air-sea interaction under the condition of a convective boundary layer. Recently, FSM has been applied to solve estimation problems in ecology, hydrology, and interdependent security analysis. For lack of space, we could not include these interesting applications.

A good working knowledge at the BS level of the standard calculus sequence, differential equations, linear algebra, and a good facility with programming constitute adequate prerequisites for a course based on this book. We have used this book as a secondary text along with parts of our earlier book (Lewis et al. 2006) in a senior/first year-graduate-level course devoted to solving static and dynamic deterministic inverse problems at the University of Oklahoma, Norman, Oklahoma, and at the University of Nevada, Reno, Nevada, USA.

We have strived to eliminate typographical errors, and we would very much appreciate hearing from readers who identify remaining errors.

Norman, OK, USA   Sivaramakrishnan Lakshmivarahan
Reno, NV, USA   John M. Lewis
Norman, OK, USA   Rafal Jabrzemski

Acknowledgments

Our interest in forward sensitivity-based approach to forecast error correction began almost a decade ago. During the development of this methodology, we have interacted and gained from the discussions with several of our colleagues and students. We wish to thank Qin Xu, National Severe Storms Laboratory, Norman, Oklahoma, for his constant support; he has been the sounding board who steered us in the right direction with a multitude of questions and comments. We are deeply indebted to Bill Stockwell (formerly of Desert Research Institute) of Howard University for many hours of discussions relating to the role of parameter sensitivity analysis in reaction kinetics and atmospheric chemistry. We (SL) wish to record our sincere thanks to Elaine Spiller and Adam Mallen, Marquette University, and Amit Apte, International Center for Theoretical Studies (ICTS) Bangalore, for their interest and comments on the earlier versions of the results in Chap. 7. We have benefitted immensely from the questions and comments from the seminar participants at the National Center for Atmospheric Research (NCAR), Marquette University; TIFR, the Center for Applicable Mathematics, Bangalore, India; the Center for Atmospheric and Ocean Studies (CAOS), Indian Institute of Science, Bangalore, India; the Meteorological Research Center (MRI), Tsukuba, Japan; the Desert Research Institute (DRI), Reno, Nevada, USA; and Texas A&M University, College Station, Texas, USA. We are grateful to Jeff Anderson (NCAR); Amit Apte (ICTS); Mythily Ramaswamy (TIFR); M. Vanninathan (TIFR); A.S. Vasudeva Murthy (TIFR); Ravi Nanjundiah, J. Srinivasan, S.K. Satheesh, and Ashwin Seshadri (CAOS and Divecha Center for Climate Change, IISc, Bangalore, India); Elaine Spiller (Marquette University); K. Saito (MRI); and I. Szunyogh (Texas A&M) for all their interest, help, and hospitality. We wish to record our sincere appreciation for Yiqi Luo and his team at the ECO Lab, University of Oklahoma (OU), for providing the data and the impetus to apply FSM-based method to assimilate data into the terrestrial carbon ecosystem (TECO) model. Our thanks are to Randy Kolar, Kendra Dersback, and Baxter Vieux, Civil Engineering, OU, and Tom Landers and Kash Barker, Industrial and Systems Engineering, OU, for encouraging their PhD students to apply FSM-based approach to estimation problems in hydrology and interdependent risk analysis.

We are particularly grateful to several graduate students at the University of Oklahoma for using FSM in their MS thesis and PhD dissertations. Phan (2011) worked on applying FSM to assimilate data into a class of carbon sequestration models called TECO model for his MS thesis. As a part of their PhD dissertations, Tromble (2011) used FSM to estimate parameters to analyze food inundation using the ADCIRC hydrodynamic model, Looper (2013) used FSM to calibrate a class of distributed hydrologic model, and Pant (2012) used FSM to estimate parameters in a class of dynamic risk input-output model.

Junjun Hu, Humberto A. Vergara, and Hongcheng Qi offered several suggestions to improve the presentations in Part I of this book. Parts of the book were used in the course Scientific Computing-I during the fall of 2015 at OU. We thank all the students in this class for their enthusiasm, especially Blake James for his comments on Part I.

The authors wish to record their thanks to the management team in the National Severe Storms Laboratory, Norman, Oklahoma, for all their support and encouragement and at the National Oceanic and Atmospheric Administration (NOAA) management, in particular Patrina Gregory and Hector Benitez, for crafting a contract that is faithful to legal requirements for copyright of work by a US federal government employee (JL). In addition, authors (RJ) would also like to express their thanks for support from the Oklahoma Mesonet.

It has been a pleasure to work with the Springer team, especially to Ron Doering—thank you all for the interest, help, and understanding right from the inception.

Contents

Part I
Introduction to Forward Sensitivity Method

Chapter 1
Introduction

1.1 Predictability Limits

As neophytes in science, we wondered about predictability. On the one hand, we knew that an eclipse of the sun could be predicted with great accuracy years in advance; yet, some of us wondered why the 2–3 day forecasts of rain or snow were so suspect. Even before we took university courses on the subject, we could speculate, and often incorrectly, as to the reasons that underlie marked difference in predictability of these events.

In this regard, it is instructive to examine the thoughts and ideas of one of the twentieth century's most celebrated contributors to the theory of dynamical-system predictability—Edward Lorenz. He was drawn to mathematics as a young man and was headed for a Ph.D. in mathematics at Harvard after completing his M.A. degree under the direction of George Birkoff in 1940 (Fig. 1.1).

WW II interrupted his career plan. With the likelihood of being drafted into military service, he opted for training as a weather forecaster in the U.S. Army Air Corps. Upon completion of the 9-month training in the so-called Cadet program at MIT, he served as a weather officer in the Pacific Theatre. At war's end and with honorable discharge from the Army, he contemplated his future—continuation for the doctorate in mathematics at Harvard or a change of direction toward meteorology. After a long conversation with Professor Henry Houghton, chair of the meteorology department at MIT, he decided to enroll as a graduate student in meteorology at MIT. His view of weather prediction in 1947 follows:

> ...not knowing about chaos and those things then [late 1940s][1], I had the feeling that this [weather forecasting] was a problem that could be solved, but the reason we didn't make perfect forecasts was that they hadn't mastered the technique yet.
> (Lorenz, 2002, personal communication)

---

[1]The authors have inserted bracketed information in the quotations.

© Springer International Publishing Switzerland 2017
S. Lakshmivarahan et al., *Forecast Error Correction using Dynamic Data Assimilation*, Springer Atmospheric Sciences, DOI 10.1007/978-3-319-39997-3_1

Fig. 1.1 Edward Lorenz is
shown in a kimono at the First
Conference on Numerical
Weather Prediction (Tokyo,
Japan 1960) (Courtesy of
G.W. Platzman)

And Lorenz was referring to the subjective method of forecasting where experience and synoptic typing—categorization of flow structures—were the basis of forecasting. But shortly after he entered graduate school at MIT, the first numerical weather prediction (NWP) experiments took place at Princeton's Institute of Advanced study under the direction of Jule Charney. And indeed, by 1950, Charney and his team succeeded in making two successful 24-h forecasts of the transient features of the large-scale flow using a simplified dynamical model. The success set the meteorological world abuzz where speculation ran the gamut from unbridled optimism to skepticism. By the mid-1950s, operational prediction of these large-scale waves took place in both Sweden and the USA. But identification and correction of systematic errors became the major theme of operational forecasting during the late 1950s into the early 1960s[2]. Among the most obvious problems was the unrepresentativeness of observations used to initialize the models—for example, the presence of only single upper-air observations over the oceanic regions bounding the USA—from ship *Papa* (approximately 400 km south of Adak, Alaska) and ship *Charlie* (between Greenland and Iceland). Nevertheless, there was hopefulness that a more representative set of observations would go far to improve the forecasts and extend the useful range of prediction. But by the early 1960s, this hopefulness was dealt a severe blow.

As reviewed in Lorenz's scientific biography *The Essence of Chaos* (Lorenz 1995), he discovered that inadvertent introduction of minor truncation error into a low-order nonlinear model of the atmosphere led to exponential growth of the error to such a degree that the noise (error) overpowered the atmospheric signal and rendered the simulated forecast useless within a relatively short period of time. This experience led him to question the feasibility of extended-range weather prediction

---

[2]See Lewis (2005) for details on the first years of operational NWP.

(forecasts on the order of weeks to months). To solidify his conjecture on the fate of long-range weather forecasting, he searched for a more realistic nonlinear fluid-dynamical model—and he found it in 1962 while visiting his colleague Barry Saltzman at Travelers Research in Hartford, Connecticut. The model was a 2-d simulation of convection—roll convection between two parallel plates—published under the title *Finite Amplitude Convection as an Initial Value Problem-I* (Saltzman 1962). Using this simple yet non-trivial model, Lorenz was able to demonstrate the extreme sensitivity of the solution to uncertainty in the initial conditions. These small perturbations (under a specific set of choices for the non-dimensional parameters of the problem) yielded drastically different evolutions of the system—later referred to as chaotic regimes. Other choices of the parameters led to periodic and non-chaotic systems—very predictable. Lorenz's seminal contribution to the theory of predictability stemmed from this investigation as found in *Deterministic Nonperiodic Flow* (Lorenz 1963).

The chaotic systems have come to be called "unstable systems"—systems unforgiving of inaccuracy in initial conditions. Those systems that are forgiving of modest uncertainty in initial conditions are referred to as "stable systems". It is the basis for understanding the very accurate prediction of solar eclipses, for example, and the imprecise prediction of rainfall. As we look at the contributions to the predictability question in meteorology, there is a preponderance of papers dealing with forecast uncertainty in response to uncertainty in the control vector (initial conditions, boundary conditions, and parameterizations). Little attention has been paid to errors in prediction that result from physical deficiencies in the models—inexact parameterizations for example. We consider this question in Chap. 6.

Determination of predictability limits from realistic atmospheric model simulations came in the mid- to late-1960s. The plans for extended-range prediction (the order of weeks) came with the goals of the Global Atmospheric Research Program (GARP), and this dictated predictability tests with the existing general circulation (GC) models (models at Geophysical Fluid Dynamics Laboratory (GFDL), Lawrence Livermore Laboratory (LLL), and University of California Los Angeles (UCLA)). The growth rate of error was found by creating an analogue pair—the control or base state and the forecast that stemmed from a perturbation to control (to the base state initial condition). Divergence of this analogue pair produced the growth rate. The doubling time for error was 5 days when averaged over the three models (Charney 1969). These models produced results that were better than climatology out to about 2 weeks. Akio Arakawa, one of the modelers involved in the study, remembers the situation as follows:

> The report (Charney 1966, 1969)[GARP Topics, Bull. Amer. Meteor. Soc.] represents one of the major steps toward the planning of GARP. It showed, for the first time, using realistic models of the atmosphere, the existence of a deterministic predictability limit the order of 2 weeks. The report says that the limit is strictly 2 weeks, which became a matter of controversy later. To me, there is no reason that it is a fixed number. It should depend on many factors, such as the part of the space/time spectrum, climate and weather regimes, region of the globe and height in the vertical, season, etc. The important indication of

the report is that the limit is not likely to be the order of days or the order of months for
deterministic prediction of middle-latitude synoptic disturbances.
A. Arakawa, quote found in Lewis (2005)

One of the most impressive and influential studies of predictability limit for weather
prediction models came with Lorenz's study in the early 1980s that made use
of operational NWP products at the European Centre for Medium-range Weather
Forecasts (ECMWF) (Lorenz 1982). He found good analogue pairs—forecasts
starting a day apart and verifying at the same time. He used the 24-h global forecast
of geopotential at 500 mbar as one member of the pair and the verifying analysis
of this forecast as the other member. The operational forecast products for nearly
the entire winter of 1980–1981 served as the basis for calculating the divergence of
analogue pairs from the ESMWF. He found the doubling time of error (the r.m.s.
difference in the analogue pairs) to be 2.5 days. The predictability limit was found
to be 2 weeks. An excellent review of Lorenz's work and subsequent tests of the
limit at ECMWF are found in Simmons and Hollingsworth (2002).

1.2   Sources of Error that Limit Predictability

We have already mentioned the unrepresentativeness of observations as a source
of error in specifying the initial conditions of a dynamical model that in turn
contribute to the model's forecast error. Using the 2.5-day doubling time for error
as found by Lorenz (1982), the initial error increases by a factor of 4 by day 5
and by a factor of 16 by day 10. Thus, an initial R.M.S. error of $0.2\,^\circ$C turns
into an error of $0.8\,^\circ$C by day 5 and $3.2\,^\circ$C by day 10. As mentioned earlier,
immediately after the advent of the operational NWP, research centered on the
sources of numerical errors took center stage. In the immediate aftermath of
operational NWP, investigations of these errors took center stage. A pedagogical
review of the issues surrounding these types of errors is found in Richtmyer and
Morton (1967). Finding the source of "noodling" in NWP models—unstable short
waves without physical meaning—proved to be one of the most difficult problems
faced by modelers. Nearly 5 years after the advent of operational NWP, Phillips
(1959) uncovered the mystery—aliasing errors in response to nonlinear interaction
between meteorological waves. Waves were created that could not be resolved
by the model. Smoothing was suggested as a means for removing the instability
that came from these short waves, but Arakawa designed finite-difference schemes
that were true to the conservation principles of the model and thereby removed
instability and permitted long-term integration without need of artificial smoothing
(Arakawa 1966; Lilly 1997). A much more subtle—difficult to determine—source
of error in dynamical model prediction is absence or improper account for important
physical processes. For example, the turbulent transfer of heat and water vapor at
the sea-air interface is most important for short-range forecasts and climatological
long-term forecasts. Yet, it is a turbulent process in the atmosphere and the exact

quantitative expression for the process escapes us. Further, proper account for the change of phase of water substance in condensation and evaporation involves an unbelievable wide range of spatial and temporal scales that makes it difficult to capture its quantitative form. In short, cloud formation and dissipation, so essential to forecasting, is poorly parameterized. And in this age of emphasis on prediction of small-scale phenomena—environments of severe storms and flow in mountainous terrain—there is a need for boundary conditions on the nested small-scale grids. Proper account for transport of constituents between the large- and small- domains is fraught with complications. Thus, the uncertainty in the elements of control: the initial condition, the parameterizations, and boundary conditions, when coupled with an imperfect observational system and numerical approximations lead to forecast error. Despite these ubiquitous sources of error, there is proof that weather predictions have improved steadily over the past half century due in large part to the global observations from satellites. An impressive summary is found in Simmons and Hollingsworth (2002) that compares the predictability limits that came with Lorenz's monumental study at ECMWF in the early 1980s and the latest operational predictions. Most obvious is the extension of the predictability limit to 2 weeks in the southern hemisphere in response to the availability of more and better satellite observations over the ocean dominated hemisphere.

1.3   Data Assimilation in Service to Prediction

Data Assimilation is a somewhat confusing term, due in part to the various strategies (and associated labels) that serve to connect models with data. The confusion can be ameliorated to a degree if we first focus on the seminal contribution that defined the method—Gauss's work in 1801 that described his mathematical method of predicting the time and place of reappearance of a celestial object after it came into conjunction with the sun (later identified as an asteroid and Ceres was its given name). The strategic prediction rested on the theory of least squares fit under constraint[3] A detailed account of this historical event and formulation of the mathematical problem is found in Lewis et al. (2006). In these past 200 years, this cardinal theory has assumed a variety of forms. But fundamentally, each form is either classified as stochastic-dynamic or deterministic. Gauss's formulation was deterministic—strict belief in the governing constraint followed by minimization of the sum of squared differences between forecasts and observations. The stochastic-dynamic version assumes uncertainty in the constraint. A pedagogical development of both approaches is found in Lewis et al. (2006). In this monograph, we will adopt

---

[3]History of science and mathematics has accorded Adrian Legendre and Carl Gauss (shown in Fig. 1.2) with co-discovery of least squares fit under constraint—Legendre's contribution with application to geodesy published in 1805 and Gauss' application to astronomy published in 1809 (Gauss 1809). In Gauss's case, it was the fit of telescopic observations of the planetoid to Kepler's Laws.

Fig. 1.2 Carl Gauss and Adrien Legendre, co-discoverers of the least squares method (pen and ink drawing by the author, J. Lewis)

the deterministic method of data assimilation. Specifically, we will use two methods with strong overlap/equality: (1) four-dimensional variational method (4D-VAR), and the forward sensitivity method (FSM).

The basis of 4D-VAR follows Yoshi Sasaki's contributions to meteorological data assimilation (Sasaki 1958, 1970a,b,c). The method has been applied to operations (Lewis 1972) but it also has a natural appeal for construction of reanalysis data sets—a recently developed methodology to perform analysis on historical data sets that combine forecast products with observations. These data sets are especially valuable to researchers who have interest in particular events from the past. The FSM was developed by Lakshmivarahan and Lewis (2010) with the intention of integrating model sensitivity into the process of least squares fit of model to observations. It is relatively easy to implement compared to 4D-VAR, but it is computationally more expensive. One of the advantages of FSM compared to other methods of data assimilation is that it offers guidance on placement of observations that give better corrections to control. In fact, if the data assimilation methodology is unable to reach a minimum of the cost function, FSM can often provide reasons based on inadequacy of observation placement.

We review Sasaki's work and carefully develop the theory underlying FSM since it is less well known to the modeling and data assimilation community. We also introduce the reader to Pontryagin's minimization principle, a deterministic least squares assimilation method that has been a mainstay in control theory—especially as a tool that can make adjustments to control that govern the path of rockets and other space-vehicles. This minimization principle has been applied to meteorology (Lakshmivarahan et al. 2013) and it offers a strategy for uncovering systematic error in models—in large part due to poorly parametrized physical process.

1.4 Overview and Goals

As we have seen in our historical review, knowledge of model predictability and its limit is central to understanding the propagation of model error. Are we dealing with an unstable or stable system, and what is the doubling time of error? Through numerical experiment we can estimate this doubling time. And then, application of data assimilation serves to correct the model error. In Part I, theory of data assimilation is presented—4D-VAR, FSM, and Pontryagin's minimization principle. At various junctures in the presentation, a variety of problems are introduced that demonstrate the power of correcting forecast errors by 4D-VAR and FSM. Among these problems are the following: (1) Poincaré's restricted 3-body problem, a classic unstable system still undergoing investigation by scholars interested in Chaos, (2) Gauss's stable problem of finding the path of Ceres, and (3) the stable dynamics of oceanic inertia oscillations under the influence of friction, as studied by oceanographers Gustafson and Kullenberg.

In Part II, we apply our methods to real-world problems that we have faced in our own research. Specifically, we examine the problem of air mass modification over the Gulf of Mexico—a problem that still presents challenges to severe storm forecasters. We then examine the "shallow water" problem, a dynamical system that has seen wide application in the study of gravity wave propagation associated with tsunamis and seiches. Our goal is to offer guidance to researchers that need to correct model error. We have viewed the problem from a macroscopic perspective where predictability and its limit is central to devising a correction strategy. And this strategy rests on deterministic data assimilation as found in 4D-VAR, FSM, and Pontryagin's minimum principle.

1.5 A Classification of Forecast Dynamics and Errors

In this section we start by introducing the basic notation (also refer to Appendix A at the end of the book) and provide a useful classification of forecast errors.

1.5.1 Forecast Dynamics

Let $\mathbf{x}(t) \in \mathbb{R}^n$ denote the state and let $\boldsymbol{\alpha} \in \mathbb{R}^p$ denote the parameters of a deterministic dynamical system, where $\mathbf{x}(t) = (x_1(t), x_2(t), \ldots, x_n(t))^T$ and $\boldsymbol{\alpha} = (\alpha_1(t), \alpha_2(t), \ldots, \alpha_p(t))^T$ are column vectors, n and p are positive integers, $t \geq 0$ denotes the time, and superscript T denotes the transpose of the vector or matrix. Let $f : \mathbb{R}^n \times \mathbb{R}^p \times \mathbb{R} \to \mathbb{R}^n$ be a mapping, where $\mathbf{f}(\mathbf{x}, \boldsymbol{\alpha}, t) = (f_1, f_2, \ldots, f_n)^T$ with $f_i = f_i(\mathbf{x}, \alpha, t)$ for $1 \leq i \leq n$. The vector spaces $\mathbb{R}^n$ and $\mathbb{R}^p$ are called the model space and parameter space, respectively.

Consider a dynamical system described by a system of ordinary nonlinear differential equations of the form

$$\frac{d\mathbf{x}}{dt} = f(\mathbf{x}, \boldsymbol{\alpha}, t) \tag{1.5.1}$$

or in component form

$$\frac{dx_i}{dt} = f_i(\mathbf{x}, \alpha, t) \tag{1.5.2}$$

where $d\mathbf{x}/dt$ denotes the time derivative of the state $\mathbf{x}(t)$, with $\mathbf{x}(0) \in \mathbb{R}^n$ the given initial condition. The control vector for the model is given by $\mathbf{c} = (\mathbf{x}(0), \boldsymbol{\alpha}) \in \mathbb{R}^n \times \mathbb{R}^p$, the combination of initial condition and parameters referred to as the control space. It is tacitly assumed that the map of $\mathbf{f}$ in (1.5.1) and (1.5.2) is such that the solution $\mathbf{x}(t) = \mathbf{x}(t, x(0), \alpha) = \mathbf{x}(t, \mathbf{c})$ exists and it is further assumed that $\mathbf{x}(t)$ has a smooth dependence on the control vector $\mathbf{c}$ such that the first $k(\geq 1)$ partial derivatives of $\mathbf{x}(t)$ with respect to the components of $\mathbf{c}$ also exist. The solution $\mathbf{x}(t)$ of (1.5.1) is known as the deterministic forecast of the state of the system at time $t > 0$. If the map $\mathbf{f}(\cdot)$ in (1.5.1) and (1.5.2) depends explicitly on t, then this system is called a time varying or nonautonomous system; if $\mathbf{f}(\cdot)$ does not depend on t, then the system is known as a time invariant or autonomous system.

Let $\mathbf{z}(t) \in \mathbb{R}^m$ be the observation vector obtained from the field measurements at time $t \geq 0$. Let $\mathbf{h} : \mathbb{R}^n \to \mathbb{R}^m$ be the mapping from the model space $\mathbb{R}^n$ to the observation space $\mathbb{R}^m$.

If $\hat{\mathbf{x}}(t)$ denotes the (unknown) true state, then we assume that $\mathbf{z}(t)$ is given by

$$\mathbf{z}(t) = \mathbf{h}(\hat{\mathbf{x}}(t) + \nu(t), \tag{1.5.3}$$

where $\nu(t) \in \mathbb{R}^m$ is the additive (unobservable and unavoidable) noise. The mapping $\mathbf{h}(\cdot)$ is known as the forward operator or the observation operator. It is further assumed that $\nu(t)$ is a white Gaussian noise with mean zero possessing a known covariance matrix $\mathbf{R}(t) \in \mathbb{R}^{m \times m}$. That is, $\nu(t) \sim N(0, \mathbf{R}(t))$.

1.5.2 Forecast Errors

The forecast error $\mathbf{e}_p(t) \in \mathbb{R}^m$ is defined as follows:

$$\mathbf{e}_p(t) \equiv \mathbf{z}(t) - \mathbf{h}(\mathbf{x}(t)) = \mathbf{b}(t) + \nu(t), \tag{1.5.4}$$

the sum of a deterministic part

$$\mathbf{b}(t) = \mathbf{h}(\hat{\mathbf{x}}(t)) - \mathbf{h}(\mathbf{x}(t)), \tag{1.5.5}$$

and the random part $v(t)$ induced by the observation noise. Our immediate goal is to analyze and isolate sources and types of forecast errors.

First, if the model map $f(\cdot)$ and the forecast operator $h(\cdot)$ are without error, that is, exact, and if the control vector c is also error free, then the deterministic forecast $x(t)$ must be correct in the sense that $x(t) = \hat{x}(t)$, the true state. Then from (1.5.4), the forecast error is purely random or white Gaussian noise (error of first kind). That is

$$e_p(t) = v(t). \tag{1.5.6}$$

If $f(\cdot)$ and $h(\cdot)$ are known perfectly but c has an error, then the forecast $x(t)$ will have a deterministic error induced by the incorrect control vector (error of the second kind). In such case, we can, in principal, decompose the forecast error as a sum

$$e_p(t) = b(c, t) + v(t). \tag{1.5.7}$$

where the deterministic part $b(c, t) = h(\hat{x}(t)) - (x(t))$ is purely a function of the error in the control vector. Finally, if f, h and c are all in error, then the forecast error e_F is given by

$$e_F = b(c, f, h, t) + v(t) \tag{1.5.8}$$

where the deterministic part represents the resulting combined errors in f, h and c (error of the third kind).

1.6 Organization of the Monograph

The book is divided into two parts. Part I develops the theory of forecast error corrections and Part II deals with applications. Correcting errors of the first kind is well established and has hinged on the work of pioneers like Carl Gauss and Ronald Fisher. In Chaps. 2–4 we develop the theory for correcting the forecast error of the second kind using the forward sensitivity method (FSM). The more difficult case of correcting forecast errors of the third kind using the celebrated Pontryagin's Minimum Principle (PMP) is covered in Chap. 5. Part II contains two chapters. In Chap. 6 we demonstrate the power of FSM method by applying it to the return-flow problem over the Gulf of Mexico. dealing with the Lagrangian flow of cold air over the warm waters of the Gulf of Mexico. The last chapter (Chap. 7) again deals with the application FSM to another version of the Lagrangian flow described by the linear shallow water model.

1.7   Exercises

1.7.1   Demonstrations and Problems

Demonstration: Population Dynamics

Background: The U.S. Census Bureau conducts a survey of the country's population every 10 years (at the start of decades) as mandated by the U.S. Constitution, Article I, Sect. 2. The first U.S. census was conducted in 1790 and led to a count of 3.9 million (3.9 M) people. The 2010 census (the 23rd census) counted 309 M people, a 10 % increase over the 2000 census, and nearly an 80-fold increase in the U.S. population over 220 years. The primary motivation for the census is to apportion seats in Congress (the House of Representatives) for each state in accord with the population of each state. California has the most seats in the House of Representatives (52) while Alaska has the fewest (1). There are 435 seats in the House.

The statistics on population during the several decades following the first census are given in Table PD.1.

Evolution of the population follows the exponential model where increase in population over a certain interval of time is proportional to the population that existed at the start of that interval. The governing equation takes the form

$$\frac{dP(t)}{dt} = \lambda P(t), \text{ where } P(0) = P(t = 0), \tag{PD.1}$$

where P is population and $\lambda > 0$ is a proportionality constant.

The evolution of population is given by

$$P(t) = P(0)e^{\lambda t}$$

where a value of $\lambda = 0.30$ per decade reasonably well represents the population growth during the first several decades of the census. Thus, values of t in Table PD.1 are the integers 1,2,... (shown in the first column) representative of decades.

Problem PD.1. Define analogues as the population forecast from 1790 to 1800 and the "analysis" (the census at 1800). Find the doubling time of the initial error (difference between the "analysis" at 1800 and the forecast from 1790 to 1800). Plot the forecast to visually display the increase error for this unstable system.

Table PD.1 Census data
(1790–1830)

Decade	Year	Count
0	1790	3.9 M
1	1800	5.3 M
2	1810	7.2 M
3	1820	9.6 M
4	1830	12.8 M

Solution:
Forecast to 1800: 5.2644 M, Analysis at 1800: 5.30 M,
Initial Error: 5.30 - 5.2644 = 0.0356 M (= 35,600).
    Evolution of the difference (D)

$$D = 0.00356\, e^{0.3t}.$$

Solve for t when $D = 2(0.0356)$ $t = \ln 2/0.3 = 2.31$ decades or 23.1 years.

Demonstration: Burgers' Laminar-Turbulent Model

J.M. Burgers was a Dutch hydrodynamicist who explored turbulent flow with simplified dynamical model that brought understanding to experiments with turbulent motion such as Osborne Reynolds classic experiment with flow in long narrow pipes. In his paper Mathematical Examples Illustrating Relations Occurring in Theory of Turbulent Fluid Motion (Burgers 1939), he introduced several of these dynamical models. The one we choose to study is labeled "First Preliminary Example" (Burgers 1939, pp. 9–12). Here he is interested in the interplay between laminar and turbulent flow. The governing equations are

$$\frac{dU}{dt} = P - \kappa U - W^2$$

$$\frac{dW}{dt} = UW - \kappa W \tag{BLT.1}$$

where U represents laminar velocity and W represents the turbulent velocity— function of time t only, P represents an external force (such as pressure) and κ is a coefficient of friction.

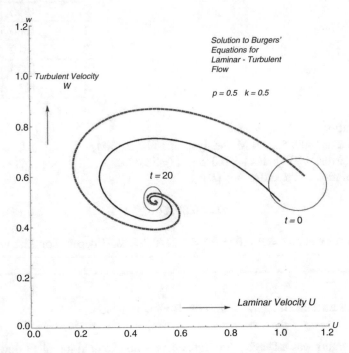

Fig. BLT.1 Trajectories of fluid in terms of the laminar and turbulent motion components

The energy equation for this flow regime is

$$\frac{\mathrm{d}}{\mathrm{d}t}\left[\frac{1}{2}(U^2 + W^2)\right] = PU - \kappa(U^2 + W^2). \qquad \text{(BLT.2)}$$

The term PU is work done by the external force and $-\kappa(U^2+W^2)$ represents the dissipation of energy in the system due to both laminar and turbulent flow.

We look at the case where $P > \kappa^2$ and in this regime there are two stable points (stable stationary solutions where the motion becomes steady)—at $U=\kappa, W = \pm\sqrt{P - \kappa^2}$—and one unstable stationary solution at $W = 0$, $U = P/\kappa$ (where a perturbation about this point diverges from the point). Figure BLT.1 shows the trajectory of two fluid parcels, one that starts at $(U = 1.0, W = 0.5)$ and the other that starts at $(U = 1.1, W = 0.6)$. Both fluid particles converge to the stable stationary point $(U = 0.5, W = 0.5)$ when the pressure $P = 0.5$ and the frictional coefficient $\kappa = 0.5$. This regime exhibits convergence of solutions.

Problem PBLT.1. In the example shown, the "distance" between initial conditions is $0.1\sqrt{2}$. Find the time required to decrease this distance by a factor of 2.

Demonstration: Restricted 3-Body Problem

In mid-to late-nineteenth century George William Hill, a mathematician with a strong interest in celestial mechanics, made an exhaustive study of the orbital dynamics of the moon in response to the action of gravitational attraction to both earth and sun. It was a classic 3-body problem that could be simplified assuming the moon exerted no effect on the motion of the sun and little effect on the motion of earth. It inspired others, especially Henri Poincaré to examine the even simpler problem of a satellite in the neighborhood of two massive heavenly bodies where the satellite exerts no influence on either of the two massive bodies. This problem became known as the restricted 3-body problem—where the dynamics of this system have been clearly derived and discussed in Keith Symon's book Mechanics (Symon 1971). We have followed Symon's development and put the governing equations in non-dimensional form.

Some simplifying assumptions beyond those mentioned above follow: (1) the two massive bodies move in a circular path about their center of gravity (the classic 2-body problem), and (2) the satellite and the massive bodies remain in the same plane. The accompanying figure displays the forces acting on the satellite (S) where the trajectory is referred to the rotating rectangular Cartesian coordinate system of the 2-body problem—that is, the rotation rate of the two heavenly bodies under the action of the mutual gravitational attraction. The x- and y-axes pass through the center of gravity (CG) of the system—the axes rotate at the rate determined by the 2-body problem dynamics and we take this rotation to be counterclockwise.

To place the problem in a practical/realistic setting, let us assume the massive bodies have mass M_1 and M_2, the heavier $M_1 = 6 \times 10^{24}$ kg (roughly the mass of earth) and the lighter body's mass is $M_2 = 2.57 \times 10^{24}$ kg. Further, let's assume the bodies are a distance L apart $= 2.86 \times 10^7$ km ($\sim 20\%$ of the astronomical Unit (A.U.)—where 1 A.U. is the nominal distance between earth and sun. On the diagram in Fig. 3BP.1, this distance is represented as 1.0 (non dimensional). From the dynamics of the two-body problem, the rotation rate is

$$\Omega^2 = \left(\frac{2\pi}{T}\right)^2 = \frac{G}{L^3}(M_1 + M_2) = \frac{GM}{L^3}, \tag{3BP.1}$$

where M is the sum of masses and G is the gravitational constant ($= 6.67 \times 10^{-11}$ s^{-2} kg^{-1} m^3).

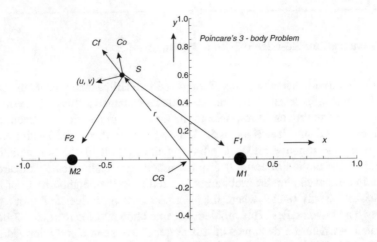

Fig. 3BP.1 Forces acting on the satellite in the presence of the two massive heavenly bodies

Thus $T \sim 40$ years. Let M_1 be located at $(x, y) = (x_1, 0)$ where $x_1 > 0$, and M_2 located at $(x, y) = (x_2, 0)$ where $x_2 < 0$, where $(x, y) = (0, 0)$ is the center of gravity (CG). Under the assumption of circular rotation of the two massive bodies about their CG, $x_1 = -\frac{M_2}{M} x_2$, and since $L = x_1 - x_2$, $x_1 = \mu L$ where $\mu = \frac{M_2}{M} = \frac{2.57}{8.57} = 0.30$ and $x_2 = (\mu - 1)L$, the masses M_1 and M_2 are located at $0.3\,L$ and $-0.7\,L$, respectively. Choosing the spatial scale to be L, the nondimensional form of x_1 and x_2, are μ and $\mu - 1$, respectively. The y-dimension is similarly scaled. The scale for time is the period of rotation, $T = 40$ years.

In the rotating coordinate system, the forces on the satellite are the gravitational attraction forces F_1 (toward M_1) and F_2 (toward M_2). The velocity of the satellite is represented by $u = \frac{dx}{dt}$ and $v = \frac{dy}{dt}$, and the acceleration $\frac{d^2 x}{dt^2}$ and $\frac{d^2 y}{dt^2}$. In the coordinate system under counterclockwise rotation, the Coriolis force (Co) is directed to the right of the velocity vector, and the centrifugal force (Cf) is directed outward along the line running from CG to the position of the satellite. The force diagram is shown in Fig. 3BP.1.

We can take the second-order ordinary differential equations (ODE's) of motion—the Newtonian equations of motion—and reconfigure them as a set of four ODE's through the action of defining a new set of variables—all non dimensional. The spatial variables are scaled by L and the velocities are scaled by L/T. The new variables are:

- $\zeta_1(t)$: x—coordinate of the satellite
- $\zeta_2(t)$: u, the satellite's velocity component in the x-direction
- $\zeta_3(t)$: y—coordinate
- $\zeta_4(t)$: v, the satellite's velocity component in the y-direction

The governing equations take the form:

$$\frac{d\zeta_1(t)}{dt} = \zeta_2(t)$$

$$\frac{d\zeta_2(t)}{dt} = -\frac{4\pi^2(1-\mu)(\zeta_1(t)-\mu)}{\left[(\zeta_1(t)-\mu)^2 + (\zeta_3(t))^2\right]^{3/2}}$$

$$-\frac{4\pi^2(\mu)(\zeta_1(t)+1-\mu)}{\left[(\zeta_1(t)+1-\mu)^2 + (\zeta_3(t))^2\right]^{3/2}}$$

$$+ 4\pi^2\zeta_1(t) + 4\pi\zeta_4(t)$$

$$\frac{d\zeta_3(t)}{dt} = \zeta_4(t)$$

$$\frac{d\zeta_4(t)}{dt} = -\frac{4\pi^2(1-\mu)\zeta_3(t)}{\left[(\zeta_1(t)-\mu)^2 + (\zeta_3(t))^2\right]^{3/2}}$$

$$-\frac{4\pi^2\mu\zeta_3(t)}{\left[(\zeta_1(t)+1-\mu)^2 + (\zeta_3(t))^2\right]^{3/2}}$$

$$+ 4\pi^2\zeta_3(t) - 4\pi\zeta_2(t)$$

(3BP.2)

The solution to these equations require four initial conditions: $\zeta_1(0)$, $\zeta_2(0)$, $\zeta_3(0)$, $\zeta_4(0)$.

For variety, we have investigated the solution of the 3-body problem for a different set of parameters than used in our basic development. In this case we choose $M_2 = 5.14 \times 10^{24}$ kg and $M_1 = 3.43 \times 10^{24}$ kg, i.e., the same value of $M = M_1 + M_2$ = 8.57×10^{24} kg as found in the earlier development. The separation is also the same, $L = 2.86 \times 10^7$ km. In this case, $\mu = 0.6$ so $x_1 = 0.6$ (nondimensional) and $x_2 = -0.4$. The rotation rate remains 40 years. The governing equations are exactly the same as in our basic demonstration, only the factor μ has changed.

We take three sets of initial conditions: slight change in the initial position of the satellite but no change in the initial velocity. Both nondimensional initial velocities are -0.1. As shown in the Fig. 3BP.2, the trajectories are quite similar as the satellite circles M_1, but by the time it passes between M_1 and M_2, the trajectories begin to differ significantly. By the nondimensional time $t = 2$ (80 years), the positions differ greatly, the order of $\sim 0.5\,L$. Results are displayed in Fig. 3BP.2. Drawings of Poincaré are displayed in Fig. 3BP.3.

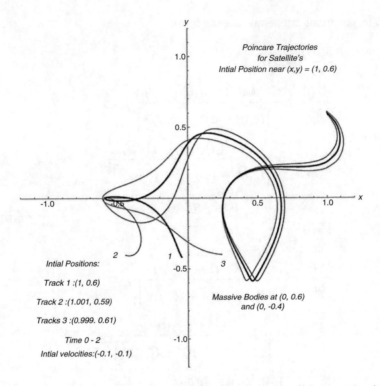

Fig. 3BP.2 Poincaré Trajectories of the satellite in response to slightly varying initial conditions

Fig. 3BP.3 Henri Poincaré in his youth and as an older man (pen and ink drawing by the author, J.Lewis)

Chapter 2
Forward Sensitivity Method: Scalar Case

In this chapter, using a simple deterministic nonlinear, scalar model, we derive the dynamics of evolution of the first-order and second-order forward sensitivities of the solution or model forecast with respect to control (the initial condition and parameters). It is assumed that the given model is perfect and the forecast errors, if any, are due only to the errors in the control consisting of the initial condition and the parameters. Refer to Sect. 1.5 for a discussion on classification of forecast errors. Given a set of N noisy observations of the phenomenon that the given model is expected to capture, we introduce a new data assimilation strategy called the forward sensitivity method (FSM). As the name implies, this strategy relies on evolution of model sensitivities—the changes in forecast as a function of changes in the control vector. These sensitivities march forward in time in step with the model forecast.

In Sect. 2.1 we derive equations governing the evolution of the first-order forward sensitivities. The basis for the FSM is developed in Sect. 2.2. The equations governing evolution of the second-order forward sensitivities are derived in Sect. 2.3. Discrete time analogs of the dynamics of evolution for first-order and second-order forward sensitivities are given in Sect. 2.4. A class of second-order methods for deterministic dynamic data assimilation using the second-order forward sensitivity is developed in Sect. 2.5. The last Sect. 2.6 contains a discussion of the computation of the Lyapunov index which is the long term average rate of growth of small errors in the initial conditions which in turn depends on the evolution of the first-order forward sensitivity.

© Springer International Publishing Switzerland 2017
S. Lakshmivarahan et al., *Forecast Error Correction using Dynamic Data Assimilation*, Springer Atmospheric Sciences, DOI 10.1007/978-3-319-39997-3_2

2.1   Evolution of First-Order Sensitivities

Let $f : \mathbb{R} \times \mathbb{R} \to \mathbb{R}$ and consider a continuous time dynamical system described by a scalar, non-linear, first-order, autonomous, ordinary differential equation of the type

$$\dot{x} = f(x, \alpha), \tag{2.1.1}$$

with $x(0) = x_0 \in \mathbb{R}$ as the initial condition and $\alpha \in \mathbb{R}$ as a real parameter. It is assumed that $f(x, \alpha)$ satisfies the conditions necessary (such as Lipschitz condition) for the existence and uniqueness of the solution that continuously depends on the initial condition x_0, parameter α and the time variable t. Thus, the solution

$$x(t) = x(t, x(0), \alpha) = x(t, \mathbf{c}), \tag{2.1.2}$$

where $\mathbf{c} = (x(0), \alpha)^T \in \mathbb{R}^2$ is called the control vector. The partial derivatives $\partial x(t)/\partial x(0)$ and $\partial x(t)/\partial \alpha$ are respectively the sensitivities of the solution $x(t)$ at time $t(> 0)$ with respect to x_0 and α. Hence these derivatives are called forward sensitivities. Likewise, the second partial derivatives $\partial^2 x(t)/\partial x^2(0)$, $\partial^2 x(t)/\partial \alpha^2$ and $\partial^2 x(t)/\partial x(0)\partial \alpha$ are known as the second-order forward sensitivities. Our goal in this section is to derive the dynamics of evolution of the first-order forward sensitivities.

To this end, taking the partial derivatives of both sides of (2.1.1) with respect to the initial condition $x(0)$, we get

$$\frac{\partial \dot{x}}{\partial x(0)} = \frac{d}{dt} \left(\frac{\partial x(t)}{\partial x(0)} \right) = \left(\frac{\partial f(x, \alpha)}{\partial x(t)} \right) \left(\frac{\partial x(t)}{\partial x(0)} \right). \tag{2.1.3}$$

To simplify the notation, define

$$u_1(t) = \frac{\partial x(t)}{\partial x(0)}, \quad D_{x(t)}(f) = \frac{\partial f(x, \alpha)}{\partial x}, \tag{2.1.4}$$

where subscript 1 refers to the first-order sensitivity.

Using these definitions (2.1.3) becomes

$$\dot{u}_1(t) = D_{x(t)}(f)u_1(t), \tag{2.1.5}$$

which is a scalar, linear, first-order, homogeneous, non-autonomous, ordinary differential equation that describes the evolution of the first-order sensitivity $u_1(t)$ when $D_x(f)$ is the Jacobian of f that varies along the trajectory of (2.1.1). Since

$$u_1(0) = \frac{\partial x(0)}{\partial x(0)} = 1, \tag{2.1.6}$$

the initial condition for (2.1.5) is $u_1(0) = 1$.

The solution $u_1(t)$ of (2.1.5) is given by

$$u_1(t) = e^{\int_0^t D_{x(\tau)}(f(x(\tau),\alpha))d\tau} u_1(0). \tag{2.1.7}$$

Again differentiating (2.1.1) with respect to α we obtain

$$\frac{\partial \dot{x}}{\partial \alpha} = \frac{d}{dt}\left(\frac{\partial x(t)}{\partial \alpha}\right) = \left(\frac{\partial f}{\partial x(t)}\right)\left(\frac{\partial x(t)}{\partial \alpha}\right) + \frac{\partial f}{\partial \alpha} \tag{2.1.8}$$

as the sum of two terms. The first term on the right hand side of (2.1.8) arises from the implicit dependence of $f(x, \alpha)$ on α through $x(t)$ and the second term arises from the explicit dependence of $f(x, \alpha)$ on α. Define

$$v_1(t) = \frac{\partial x(t)}{\partial \alpha}, \quad \text{and} \quad D_\alpha(f) = \frac{\partial f(x, \alpha)}{\partial \alpha}. \tag{2.1.9}$$

Using (2.1.4) and (2.1.9), we can rewrite (2.1.8) succinctly as

$$\dot{v}_1(t) = D_{x(t)}(f)v_1(t) + D_\alpha(f), \tag{2.1.10}$$

which is again a scalar, linear, first-order, non-homogeneous, non-autonomous ordinary differential equation with a forcing term $D_\alpha(f)$, which is the Jacobian of f with respect to α that also varies along the trajectory of (2.1.1).

Thus $u_1(t)$ and $v_1(t)$ as solutions of (2.1.5) and (2.1.10) are called first-order forward sensitivity functions and are computed using the algorithm given in Algorithm 2.1.

Algorithm 2.1 Computation of the forward sensitivity function

Given $f(x, \alpha)$, the initial condition $x(0)$, the value of the parameter α, and $T > 0$.
Step 1: Compute the Jacobian $\mathbf{D}_x(f)$ and $\mathbf{D}_\alpha(f)$
Step 2: Solve (2.1.1) analytically, if possible, or numerically using some discretization scheme—Euler method, 4th-order Runge-Kutta method and compute $x(t)$ for $0 \le t \le T$
Step 3: Evaluate $\mathbf{D}_x(f)$ along the trajectory computed in Step 2 and solve (2.1.5) and compute the evolution of the forward sensitivity function $u_1(t), 0 \le t \le T$.
Step 4: Evaluate $\mathbf{D}_\alpha(f)$ along the trajectory computed in Step 2 and solve (2.1.10) and compute the evolution of the forward sensitivity function $v_1(t), 0 \le t \le T$.

The following example illustrates the computation of the first-order forward sensitivity functions.

Example 1. Consider the case where cold continental air moves out over an ocean of constant surface temperature. We follow a column of air in a Lagrangian frame; that is, the column of air moves with the prevailing low-level wind speed. Turbulent transfer of heat from the ocean to air warms the column. The governing linear dynamics is taken to be

$$\frac{dx}{dt} = \frac{C_T V}{H}(\theta - x) \tag{2.1.11}$$

where

- x : temperature of the air column (°C),
- θ : sea surface temperature (SST, °C),
- C_T : turbulent heat exchange coefficient (nondimensional),
- V : speed of air column (m s^{-1}),
- H : height of the column − the mixed layer (m)
- t : time (h).

Assuming the following scales for the physical variables,
$H \sim 150$ m, $V \sim 10$ m s^{-1}, $C_T \sim 10^{-3}$,
we get

$$\frac{C_T V}{H} \sim 0.25 \, \text{h}^{-1}.$$

Let $c = \dfrac{C_T V}{H}$, then (2.1.11) becomes

$$\frac{dx}{dt} = c\,(\theta - x) = f(x, \alpha),$$

whose analytic solution is

$$x(t, x_0, \alpha) = (x_0 - \theta)e^{-ct} + \theta \tag{2.1.12}$$

where x_0 is the initial condition and the parameters are θ and c.

There are three elements of control: initial condition, $x(0)$, boundary condition θ, and parameter, c. The Jacobians of f with respect to x and α are given by

$$D_x(f) = -c, \quad D_\theta(f) = c, \quad D_c(f) = \theta - x(t). \tag{2.1.13}$$

The evolution of the forward sensitivity with respect to the initial condition is given by

$$\frac{d}{dt}\left(\frac{\partial x(t)}{\partial x(0)}\right) = -c\left(\frac{\partial x(t)}{\partial x(0)}\right), \tag{2.1.14}$$

and that with respect to the parameters θ and c are given by

$$\left.\begin{array}{l} \dfrac{\mathrm{d}}{\mathrm{d}t}\left(\dfrac{\partial x(t)}{\partial \theta}\right) = -c\dfrac{\partial x(t)}{\partial \theta} + c, \\[4mm] \dfrac{\mathrm{d}}{\mathrm{d}t}\left(\dfrac{\partial x(t)}{\partial c}\right) = -c\dfrac{\partial x(t)}{\partial c} + \theta - x(t), \end{array}\right\} \tag{2.1.15}$$

where

$$\left[\dfrac{\partial x(t)}{\partial \theta}\right]_{t=0} = 0 = \left[\dfrac{\partial x(t)}{\partial c}\right]_{t=0}, \quad \text{and} \quad \left[\dfrac{\partial x(t)}{\partial x(0)}\right]_{t=0} = 1.$$

Either by solving (2.1.14) and (2.1.15) or by computing directly from (2.1.12), it can be verified that the required sensitivities evolve according to

$$\frac{\partial x(t)}{\partial x(0)} = e^{-ct},$$

$$\frac{\partial x(t)}{\partial \theta} = 1 - e^{-ct}, \tag{2.1.16}$$

$$\frac{\partial x(t)}{\partial c} = [\theta - x(0)]\, t e^{-ct}.$$

The plots of the solution and the three sensitivities when $x(0) = 1$, $x_s = 11$ and $c = 0.25$ are given in Figs. 2.1, 2.2, 2.3, and 2.4.

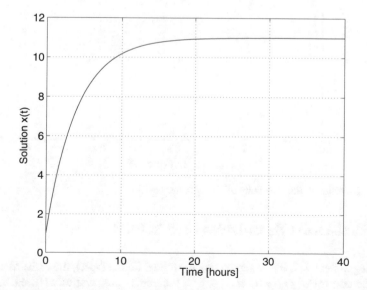

Fig. 2.1 Evolution of the solution of $x(t)$ in (2.1.12)

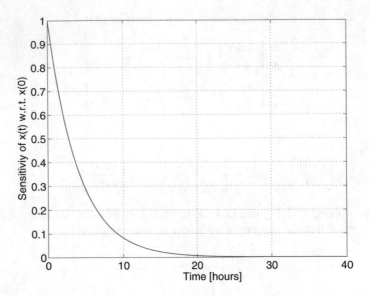

Fig. 2.2 Evolution of the sensitivity of $x(t)$ with respect to the initial condition $x(0)$

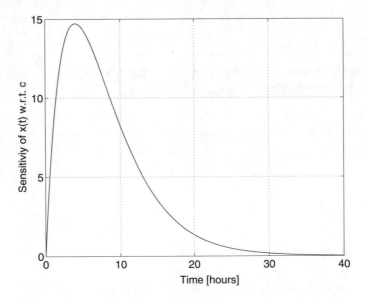

Fig. 2.3 Evolution of the sensitivity of $x(t)$ with respect to c

2.2 First-Order Sensitivities Used in FSM

Referring to Sect. 1.5, it is assumed that the model field, $f(x, \alpha)$, the forward operator $h(x)$, the control $\mathbf{c} = (x(0), \alpha)^T \in \mathbb{R}^2$, the actual observation $z(t)$ and its error covariance $R(t)$ are given. It is assumed that the current value of the control $\mathbf{c}$ is

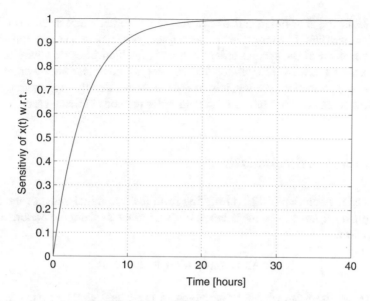

Fig. 2.4 Evolution of the sensitivity of $x(t)$ with respect to the boundary condition, θ

not exact which in turn induces error in the model forecast. Our goal is to find the correction, $\delta\mathbf{c}$ to the control vector $\mathbf{c}$, such that the new model forecast starting from $(\mathbf{c} + \delta\mathbf{c})$ will render the forecast error, $\mathbf{e}_p(t)$ purely random. That is, we wish to remove the systematic part of the forecast error induced by the erroneous control. To this end, we start by quantifying the actual change Δx in the solution $x(t) = x(t, \mathbf{c})$ resulting from a change $\delta\mathbf{c}$ in $\mathbf{c}$. Using the standard Taylor series expansion, we get

$$\Delta x = x(t, \mathbf{c} + \delta\mathbf{c}) - x(t, \mathbf{c}) = \sum_k \delta^k x, \qquad (2.2.1)$$

where the kth variation of $x(t)$ is given by

$$\delta^k x = g\left(\frac{\partial^k x}{\delta\mathbf{c}^k}, \delta\mathbf{c}\right) \qquad (2.2.2)$$

and g is a known function of the kth partial derivative of $x(t)$ with respect to $\mathbf{c}$ and the perturbation $\delta\mathbf{c}$. While using higher-order correction terms in (2.2.1) would lead to improved accuracy, inclusion of higher order terms invariably leads to a more complex inverse problem as will become evident from the following analysis. Recognizing this trade-off between complexity and accuracy, a useful compromise is to restrict the sum in (2.2.1) to the first two terms at the most, that is

$$\Delta x = \delta x + \delta^2 x, \qquad (2.2.3)$$

where δx is the first-order and $\delta^2 x$ is the second-order correction term. In the following, we first illustrate the first-order analysis, where Δx is approximated by δx. Inclusion of the second-order analysis is justified in cases where the field, $f(x, \alpha)$ and/or the forward operator $h(x)$, are highly nonlinear and warrant use of second-order correction. Further, it is shown below that iterative application of first-order method can be used in lieu of second-order methods in many cases.

2.2.1 First-Order Analysis

Let $\bar{x}(t)$ be the solution of (2.1.1) starting from the true initial value of the control $\mathbf{c} = (x(0), \alpha)^T$. Let $x(t)$ be the solution of (2.1.1) starting from the perturbed value of the control

$$\mathbf{c} + \delta \mathbf{c} = (x(0) + \delta x(0), \alpha + \delta \alpha)^T.$$

Let $\delta x(t)$ be the first-order approximation to the actual change $x(t) - \bar{x}(t)$—see Fig. 2.5. From first principles, that is, applying the first-order Taylor expansion to (2.1.2), we have

$$\Delta x \approx \delta x = D_{x(0)}(x)\delta x(0) + D_\alpha(x)\delta \alpha \tag{2.2.4}$$

where

$$\mathbf{D_{x(0)}(x)} = \frac{\partial x(t)}{\partial x(0)} = u_1(t), \ \mathbf{D_\alpha(x)} = \frac{\partial x(t)}{\partial \alpha} = v_1(t), \tag{2.2.5}$$

are the first-order sensitivities of $x(t)$ with respect to the initial condition $x(0)$ and the parameter α, respectively.

Knowing the first-order forward sensitivities, we compute the corrections $\delta x(0)$ and $\delta \alpha$ by fitting the model to a given set of noisy observations in a least squares framework.

Case 1: Single Observation

Let $z(t)$ be a given single observation. Recall that

$$z(t) = h(x(t)) + v(t), \tag{2.2.6}$$

Fig. 2.5 An illustration of an initial perturbation and its impact on the solution at time t

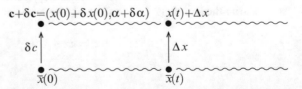

where $v(t) \sim N(0, \sigma_0^2)$. If the model counterpart to the observation is strictly the model variable, then $h(x(t)) = x(t)$. The first variation δx in $x(t)$ induces a first variation δh in $h(x(t))$ given by

$$\delta h = \mathbf{D_x}(h)\delta x, \qquad (2.2.7)$$

where

$$\mathbf{D_x}(h) = \frac{\partial h(x)}{\partial x}. \qquad (2.2.8)$$

Again, if $h(x) = x$, then $\mathbf{D}_x(h) = 1$.

Substituting (2.2.4) into (2.2.7) yields

$$\delta h = H_1(t)\delta x(0) + H_2(t)\delta\alpha, \qquad (2.2.9)$$

where

$$\begin{aligned}
H_1(t) &= \mathbf{D_x}(h)\mathbf{D_{x(0)}}(x) = \mathbf{D_x}(h)u_1(t), \\
H_2(t) &= \mathbf{D_x}(h)\mathbf{D_\alpha}(x) = \mathbf{D_x}(h)v_1(t).
\end{aligned} \qquad (2.2.10)$$

Now, define

$$\mathbf{H}(t) = [H_1(t), H_2(t)] \in \mathbb{R}^{1\times 2}, \text{ and } \boldsymbol{\zeta} = (\zeta_1, \zeta_2)^T, \qquad (2.2.11)$$

where $\boldsymbol{\zeta}_1 = \delta x(0)$, and $\boldsymbol{\zeta}_2 = \delta\alpha$. Then (2.2.9) can be succinctly written as

$$\delta h = \mathbf{H}(t)\boldsymbol{\zeta}. \qquad (2.2.12)$$

Given an operating point $\mathbf{c}$, our goal is to find the perturbation $\delta\mathbf{c} = \boldsymbol{\zeta}$ such that the given observation $z(t)$ matches its model counterpart to a first-order accuracy, that is

$$z(t) = h(x(t) + \delta x) \approx h(x(t)) + \delta h, \qquad (2.2.13)$$

Thus, the forecast error $e_F(t)$ when viewed from the observation space becomes

$$e_F(t) \equiv z(t) - h(x(t)) \approx \delta h, \qquad (2.2.14)$$

where $x(t)$ is the forecast computed from the incorrect control $\mathbf{c}$. From (2.2.12) and (2.2.14), the perturbation $\delta\mathbf{c} = \boldsymbol{\zeta}$ is required to satisfy

$$\mathbf{H}(t)\boldsymbol{\zeta} = \mathbf{e}_F(t). \qquad (2.2.15)$$

But ζ has two or more components—two if the parameter space has dimension one. Thus, ζ cannot be determined from a single observation and the associated single equation. The problem as stated is under-determined.

The approach summarized in Box 2.1 offers a least squares solution to the problem by requiring that the norm of the corrections be minimized. In short, for the case of two unknowns, the requirement that the sum of the squared adjustments be a minimum yields an expression for the optimal (least squares) adjustment.

Box 2.1 Method of Lagrangian multipliers.

Lagrangian multiplier method: under-determined case.

Let the model counterpart matrix $\mathbf{H}$: $[h_1, h_2] \in \mathbb{R}^{1 \times 2}, \mathbf{x} = (x_1, x_2)^T \in \mathbb{R}^2$ and $z \in \mathbb{R}$. Let $z = \mathbf{H}\mathbf{x}$ or $h_1 x_1 + h_2 x_2 = z$. Rewriting this as $x_1 = \frac{1}{h_1}[z - h_2 x_2]$, it follows that for each $x_2 \in \mathbb{R}$, there is a unique x_1 satisfying this relation. Hence, we are left with a problem admitting infinitely many solutions (Lewis et al. 2006, Chap. 5).

The method of Lagrangian multipliers helps to find a unique solution by seeking the one with the minimum norm (same as energy) subject to the constraint. To this end, for $\lambda \in \mathbb{R}$ define the Lagrangian

$$L(\mathbf{x}, \lambda) = \frac{1}{2}\mathbf{x}^T\mathbf{x} + \lambda(z - \mathbf{H}\mathbf{x})$$

Here λ is called the undetermined Lagrangian multiplier. The necessary conditions for the minimum of $L(x, \lambda)$ are given by

$$\nabla_x L(\mathbf{x}, \lambda) = \mathbf{x} - \mathbf{H}^T \lambda = 0$$
$$\nabla_\lambda L(\mathbf{x}, \lambda) = \mathbf{z} - \mathbf{H}\mathbf{x} = 0$$

Substituting the first equation in the second, we get

$$\lambda = \left(\mathbf{H}\mathbf{H}^T\right)^{-1} z$$

which when substituted back into the first equation yields the unique solution

$$x_{LS} = \mathbf{H}^T \left(\mathbf{H}\mathbf{H}^T\right)^{-1} z$$

$$\zeta_{LS} = \mathbf{H}^T \left(\mathbf{H}\mathbf{H}^T\right)^{-1} e_F(t), \tag{2.2.16}$$

which can be computed using the algorithm given in Algorithm 2.2.

Algorithm 2.2 Algorithm for computing the forecast error correction

---

> *Step 1*: Compute $\mathbf{D_x}(h)$
> *Step 2*: Using $\mathbf{D_{x(0)}}(x) = u(t)$ and $\mathbf{D_\alpha}(x) = v(t)$ computed in Algorithm 2.1, now assemble $H_1 = \mathbf{D_x}(h)\mathbf{D_{x(0)}}$ and $H_2 = \mathbf{D_x}(h)\mathbf{D_\alpha}(x)$ and form $\mathbf{H} = [H_1, H_2]$.
> *Step 3*: Using $x(t)$ computed in Algorithm 2.1, compute the forecast error
> $\mathbf{e_F}(t) = \mathbf{z}(t) - \mathbf{h}\,(\mathbf{x}(t))$.
> *Step 4*: Solve the under-determined linear least-squares problem using (2.2.16) (Lewis et al. 2006, Chap. 5).

Remark 2.1 (Structure of Optimal Correction).

In this case where there is only one observation, $\mathbf{H}$ is a row matrix of size 1×2 and hence

$$\mathbf{HH}^T = \mathbf{H}_1^2(t) + \mathbf{H}_2^2(t) = \|\mathbf{H}\|^2$$

Consequently we can rewrite (2.2.16) as a product of a vector and a scalar given by

$$\zeta_{LS} = \frac{\mathbf{H}^T}{\|\mathbf{H}\|}\,\frac{\mathbf{e_F}(t)}{\|\mathbf{H}\|}, \qquad (2.2.17)$$

Since the second factor on the right hand side is a scalar that is proportional to the forecast error, $e_F(t)$, it follows that the optimal least squares correction, ζ_{LS}, is proportional to the unit vector

$$\frac{\mathbf{H}^T}{\|\mathbf{H}\|} = \frac{1}{[u_1^2(t) + v_1^2(t)]^{1/2}} \begin{bmatrix} u_1(t) \\ v_1(t) \end{bmatrix}$$

whose components are uniquely determined by the forward sensitivities $u_1(t)$ and $v_1(t)$. In other words, the optimal correction to the control is uniquely determined by the normalized forward sensitivity vector and the scaled forecast error.

Case 2: Multiple Observations

We now extend the above analysis to the case when there are observations available at N different times, beyond the initial time ($t = 0$). Let $0 < t_1 < t_2 < \ldots < t_N$ be the N different times and $z(t_1), z(t_2), \ldots, z(t_N)$ be the observations at these times. If $x(t)$ is the current model forecast, then the forecast errors at time t_i, for $1 \leq i \leq N$ are given by

$$\mathbf{e_F}(t_i) = \mathbf{z}(t_i) - \mathbf{h}\,(x(t_i)). \qquad (2.2.18)$$

If δc is the initial perturbation in the control c, the sequence of N induced perturbations are given by

$$\delta x(t_i) = \mathbf{D_{x(0)}}(x(t_i))\delta x(0) + \mathbf{D}_\alpha(x(t_i))\delta\alpha, \ 1 \le i \le N. \tag{2.2.19}$$

The first variation $\delta h(i)$, the model counterpart of the observation at time t_i, is given by

$$\delta h(i) = D_{x(t_i)}(h)\delta x(t_i) \tag{2.2.20}$$

Combining these we obtain

$$\delta h(i) = H_1(t_i)\delta x(0) + H_2(t_i)\delta\alpha, \tag{2.2.21}$$

where

$$\begin{aligned} H_1(t_i) &= D_{x(t_i)}(h)D_{x(0)}\left(x(t_i)\right), \\ H_2(t_i) &= D_{x(t_i)}(h)D_\alpha\left(x(t_i)\right). \end{aligned} \tag{2.2.22}$$

Hence, at time t_i, we have

$$\mathbf{H}(t_i)\boldsymbol{\zeta} = e_F(t_i), \tag{2.2.23}$$

where

$$\begin{aligned} \mathbf{H}(t_i) &= [H_1(t_i), H_2(t_i)] \in \mathbb{R}^{1\times 2}, \\ \boldsymbol{\zeta} &= (\delta x(0), \delta c)^T \in \mathbb{R}^2. \end{aligned} \tag{2.2.24}$$

Now define an $N \times 2$ matrix $\mathbf{H}$ obtained by stacking $H(t_i)$ in rows and the vector $\mathbf{e}_F$ by stacking $e_F(t_i)$ in rows:

$$\mathbf{H} = \begin{bmatrix} H(t_1) \\ H(t_2) \\ \vdots \\ H(t_N) \end{bmatrix} \in \mathbb{R}^{N\times 2}, \ \mathbf{e}_F = \begin{bmatrix} e_F(t_1) \\ e_F(t_2) \\ \vdots \\ e_F(t_N) \end{bmatrix} \in \mathbb{R}^N. \tag{2.2.25}$$

Then, (2.2.23) becomes

$$\mathbf{H}\boldsymbol{\zeta} = \mathbf{e}_F. \tag{2.2.26}$$

Again, from Sect. 1.5 it follows that $\mathbf{e}_F$ contains a systematic deterministic error and a random observation noise. Assuming that the matrix $\mathbf{H}$ is of full rank, that is, Rank($\mathbf{H}$) $= 2$, we get two special cases. When $N = 2$, (2.2.26) can be solved uniquely by the standard method for solving linear systems. Refer to Lewis et al. (2006, Chap. 9) for details.

We now consider the overdetermined case when $N \geq 3$. In such a case, (2.2.26) is solved by minimizing the weighted sum of squared errors

$$J(\zeta) = \frac{1}{2}(\mathbf{e}_F - H(t_i)\zeta)^T \mathbf{R}^{-1}(t_i)(\mathbf{e}_F - H(t_i)\zeta), \tag{2.2.27}$$

where $\mathbf{R} \in \mathbb{R}^{N \times N}$ diagonal matrix where the ith diagonal element represents the variance of the ith observation $z(t_i)$.

By taking gradient $J(\zeta)$ with respect to ζ (refer to Appendix A for details) and equating it to zero, we readily see that the minimized ζ_{LS} is given by the solution of the linear system

$$\left(\mathbf{H}^T \mathbf{R}^{-1} \mathbf{H}\right)\zeta = \mathbf{H}^T \mathbf{R}^{-1} \mathbf{e}_F, \tag{2.2.28}$$

where the system matrix $\left(\mathbf{H}^T \mathbf{R}^{-1} \mathbf{H}\right)$ is symmetric and positive definite. It follows (refer to Appendix A for details) that the Hessian of $J(\zeta)$ in (2.2.27) is given by

$$\nabla^2 J(\zeta) = \left(\mathbf{H}^T \mathbf{R}^{-1} \mathbf{H}\right), \tag{2.2.29}$$

which is symmetric and positive definite. The least squares (LS) solution ζ_{LS} of (2.2.28) is

$$\zeta_{LS} = \left(\mathbf{H}^T \mathbf{R}^{-1} \mathbf{H}\right)^{-1} \mathbf{H}^T \mathbf{R}^{-1} \mathbf{e}_F.$$

Referring to Lewis et al. (2006, Chap. 5 and 6), it follows that the vector $\mathbf{H}\zeta_{LS}$ is the (oblique) projection of $\mathbf{e}_F$ onto the subspace spanned by the columns of $\mathbf{H}$.

2.3   Evolution of Second-Order Sensitivities

For later use, we derive equations for evolution of second-order forward sensitivities associated with the solution of (2.1.1)—$x(t) = x(t, \mathbf{c})$ where $\mathbf{c} = (x(0), \alpha)^T$. In principle there are three equations of interest to us: one for each of $\frac{\partial^2 x(t)}{\partial x^2(0)}$, $\frac{\partial^2 x(t)}{\partial \alpha^2}$ and $\frac{\partial^2 x(t)}{\partial x(0)\partial \alpha}$ which are respectively the second partial derivations of $x(t)$ with respect to $x(0), \alpha$ and the mixed partial derivative that captures the interaction between $x(0)$ and α.

2.3.1 Evolution of $\partial^2 x(t)/\partial x^2(0)$

Since we already know the evolution of the first-order sensitivity of $x(t)$ with respect to $x(0)$, we start with this equation given in (2.1.3). We reproduce this equation here for convenience:

$$\frac{d}{dt}\left(\frac{\partial x(t)}{\partial x(0)}\right) = \left(\frac{\partial f(x,\alpha)}{\partial x(t)}\right)\left(\frac{\partial x(t)}{\partial x(0)}\right). \tag{2.3.1}$$

Now, differentiate both sides with respect to $x(0)$ to get

$$\frac{\partial}{\partial x(0)}\left[\frac{d}{dt}\left(\frac{\partial x(t)}{\partial x(0)}\right)\right] = \frac{\partial}{\partial x(0)}\left[\left(\frac{\partial f(x,\alpha)}{\partial x(t)}\right)\left(\frac{\partial x(t)}{\partial x(0)}\right)\right].$$

Invoking the product rule, we get

$$\begin{aligned}
\frac{d}{dt}\left[\left(\frac{\partial^2 x(t)}{\partial x^2(0)}\right)\right] &= \left[\frac{\partial}{\partial x(0)}\left(\frac{\partial f(x,\alpha)}{\partial x(t)}\right)\right]\frac{\partial x(t)}{\partial x(0)} \\
&\quad + \left(\frac{\partial f(x,\alpha)}{\partial x(t)}\right)\frac{\partial}{\partial x(0)}\left(\frac{\partial x(t)}{\partial x(0)}\right) \\
&= \frac{\partial f^2(x,\alpha)}{\partial x^2(t)}\left(\frac{\partial x(t)}{\partial x(0)}\right)^2 + \frac{\partial f(x,\alpha)}{\partial x(t)}\left(\frac{\partial^2 x(t)}{\partial x^2(0)}\right). \tag{2.3.2}
\end{aligned}$$

Define

$$u_2 = \frac{\partial^2 x(t)}{\partial x^2(0)}, \ \mathbf{D}_x^2(f) = \frac{\partial^2 f}{\partial x^2}. \tag{2.3.3}$$

Substituting (2.3.3) in (2.3.2), the required dynamics takes the form

$$\frac{du_2}{dt} = \mathbf{D}_x(f)u_2 + \mathbf{D}_x^2(f)u_1^2 \tag{2.3.4}$$

with the initial conditions

$$u_2(0) = \left.\frac{\partial^2 x(t)}{\partial x^2(0)}\right|_{t=0} = \left.\frac{\partial}{\partial x(0)}\left(\frac{\partial x(t)}{\partial x(0)}\right)\right|_{t=0} = \frac{\partial}{\partial x(0)}(1) = 0 \tag{2.3.5}$$

which is a scalar, linear, non-autonomous, non-homogeneous, first-order ordinary differential equation.

2.3.2 Evolution of $\partial^2 x(t)/\partial \alpha^2$

Now consider the evolution of the first-order sensitivity of $x(t)$ with respect to α given in (2.1.8), which is reproduced here for convenience:

$$\frac{d}{dt}\left(\frac{\partial x(t)}{\partial \alpha}\right) = \left(\frac{\partial f(x,\alpha)}{\partial x(t)}\right)\left(\frac{\partial x(t)}{\partial \alpha}\right) + \frac{\partial f}{\partial \alpha}. \tag{2.3.6}$$

Differentiating both sides with respect to α, using product rule, we get

$$\frac{d}{dt}\left[\left(\frac{\partial^2 x(t)}{\partial \alpha^2}\right)\right] = \frac{\partial}{\partial \alpha}\left[\left(\frac{\partial f(x,\alpha)}{\partial x(t)}\right)\right]\frac{\partial x(t)}{\partial \alpha}$$

$$+ \left(\frac{\partial f(x,\alpha)}{\partial x(t)}\right)\left(\frac{\partial^2 x(t)}{\partial \alpha^2}\right) + \frac{\partial}{\partial \alpha}\left(\frac{\partial f}{\partial \alpha}\right). \tag{2.3.7}$$

While the first-order sensitivity dynamics in (2.1.3) is linear, homogeneous and non-autonomous, this second-order sensitivity equation is linear, non-homogeneous and non-autonomous.

Recall that, $f = f(x,\alpha)$ and $x(t) = x(t, x_0, \alpha)$. Hence,

$$\frac{\partial}{\partial \alpha}\left(\frac{\partial f}{\partial x}\right) = \frac{\partial^2 f}{\partial x^2}\frac{\partial x}{\partial \alpha} + \left(\frac{\partial^2 f}{\partial x \partial \alpha}\right)$$

and

$$\frac{\partial}{\partial \alpha}\left(\frac{\partial f}{\partial \alpha}\right) = \left(\frac{\partial^2 f}{\partial x \partial \alpha}\right)\frac{\partial x}{\partial \alpha} + \frac{\partial^2 f}{\partial \alpha^2}.$$

Substituting (2.3.8) in (2.3.7) and simplifying, we get

$$\frac{d}{dt}\left(\frac{\partial^2 x(t)}{\partial \alpha^2}\right) = \left(\frac{\partial^2 f}{\partial x^2(t)}\right)\left(\frac{\partial x}{\partial \alpha}\right)^2 + \left(\frac{\partial^2 f}{\partial x(t)\partial \alpha}\right)\left(\frac{\partial x(t)}{\partial \alpha}\right)$$

$$+ \left(\frac{\partial f}{\partial x(t)}\right)\left(\frac{\partial^2 x(t)}{\partial \alpha^2}\right) + \left(\frac{\partial^2 f}{\partial \alpha \partial x(t)}\right)\left(\frac{\partial x(t)}{\partial \alpha}\right)$$

$$+ \left(\frac{\partial^2 f}{\partial \alpha^2}\right). \tag{2.3.8}$$

Define

$$v_2 = \frac{\partial^2 x(t)}{\partial \alpha^2}, \qquad \mathbf{D}^2_\alpha(f) = \frac{\partial^2 f}{\partial \alpha^2}, \qquad \mathbf{D}^2_{x\alpha}(f) = \frac{\partial^2 f}{\partial x \partial \alpha}. \tag{2.3.9}$$

Substituting (2.2.5), (2.3.3) and (2.3.9) in (2.3.8), the dynamics of evolution of the second-order sensitivity of $x(t)$, with respect to α is given by

$$\frac{dv_2}{dt} = \mathbf{D}_x(f)v_2 + \mathbf{D}_x^2(f)v_1^2 + 2\mathbf{D}_{x\alpha}^2(f)v_1 + \mathbf{D}_\alpha{}^2(f), \qquad (2.3.10)$$

which is a scalar, linear, non-autonomous and non-homogeneous ordinary differential equation, with the initial condition

$$v_2(0) = \left.\frac{\partial^2 x(t)}{\partial \alpha^2}\right|_{t=0} = \left.\frac{\partial}{\partial x(0)}\left(\frac{\partial x(t)}{\partial \alpha}\right)\right|_{t=0} = \frac{\partial}{\partial \alpha}(0) = 0$$

2.3.3 Evolution of $\partial^2 x(t)/\partial\alpha\,\partial x(0)$

The evolution of this second-order cross sensitivity can be derived starting either from (2.3.1) or (2.3.6). For definiteness, we start with (2.3.6). Now differentiate both sides of (2.3.6) with respect to $x(0)$, we obtain

$$\begin{aligned}
\frac{d}{dt}\left(\frac{\partial^2 x(t)}{\partial x(0)\partial\alpha}\right) &= \left[\frac{\partial}{\partial x(0)}\left(\frac{\partial f}{\partial x(t)}\right)\right]\left(\frac{\partial x(t)}{\partial \alpha}\right) \\
&\quad + \left(\frac{\partial f}{\partial x(t)}\right)\frac{\partial}{\partial x(0)}\left(\frac{\partial x(t)}{\partial \alpha}\right) + \frac{\partial}{\partial x(0)}\left(\frac{\partial f}{\partial \alpha}\right) \\
&= \left(\frac{\partial^2 f}{\partial x^2(t)}\right)\left(\frac{\partial x(t)}{\partial x(0)}\right)\left(\frac{\partial x(t)}{\partial \alpha}\right) \\
&\quad + \left(\frac{\partial f}{\partial x(t)}\right)\left(\frac{\partial^2 x(t)}{\partial x(0)\partial\alpha}\right) + \left(\frac{\partial^2 f}{\partial\alpha\,\partial x}\right)\frac{\partial x(t)}{\partial x(0)} \qquad (2.3.11)
\end{aligned}$$

Using (2.2.5), (2.3.3) and (2.3.9) in (2.3.11), the required dynamics becomes

$$\frac{d}{dt}\left(\frac{\partial^2 x(t)}{\partial x(0)\partial\alpha}\right) = \mathbf{D}_x(f)\left(\frac{\partial^2 x(t)}{\partial x(0)\partial\alpha}\right) + \mathbf{D}_x^2(f)u_1v_1 + \mathbf{D}_{x\alpha}^2(f)u_1, \qquad (2.3.12)$$

where the initial condition

$$\left.\frac{\partial^2 x(t)}{\partial x(0)\partial\alpha}\right|_{t=0} = \left.\frac{\partial}{\partial \alpha}\left(\frac{\partial x(t)}{\partial x(0)}\right)\right|_{t=0} = \frac{\partial}{\partial x(0)}(1) = 0. \qquad (2.3.13)$$

2.4   Data Assimilation Using FSM: A Second-Order Method

Referring to (2.2.3), a second-order approximation to Δx given by

$$\Delta x \approx \delta x(k) + \delta^2 x(k) \tag{2.4.1}$$

where, as before

$$\delta x(k) = \left[\frac{\partial x(k)}{\partial x(0)}, \frac{\partial x(k)}{\partial \alpha} \right] \begin{bmatrix} \delta x(0) \\ \delta \alpha \end{bmatrix} = \mathbf{D}_c(x)\zeta \tag{2.4.2}$$

and $\mathbf{D}_c$ is the Jacobian of $x(k)$ with respect to ζ. Similarly

$$\delta^2 x(k) = \frac{1}{2} \left(\delta x(0), \delta \alpha \right) \begin{bmatrix} \frac{\partial^2 x(k)}{\partial x^2(0)} & \frac{\partial^2 x(k)}{\partial \alpha \partial x(0)} \\ \frac{\partial^2 x(k)}{\partial x(0)\partial \alpha} & \frac{\partial^2 x(k)}{\partial \alpha^2} \end{bmatrix} \begin{bmatrix} \delta x(0) \\ \delta \alpha \end{bmatrix}$$

$$= \frac{1}{2}\zeta^T \mathbf{D}_c^2(x)\zeta \tag{2.4.3}$$

where $\mathbf{D}_c^2(x)$ is the Hessian of x with respect to $\mathbf{c}$. For simplicity in algebra, again consider case of a single observation $z(k)$ at time k where

$$z(k) = h\left(x(k)\right) + v(k) \tag{2.4.4}$$

Again, from first principles, the second-order change Δh in $h(x)$ induced by the change Δx in x is given by

$$\Delta h = h\left(x + \Delta x\right) - h(x) = \delta h + \delta^2 h \tag{2.4.5}$$

where

$$\delta \mathbf{h} = \left(\frac{\partial h}{\partial x} \right) \delta x = \mathbf{D}_x(h)\delta x \text{ and}$$

$$\delta^2 h = \frac{1}{2} \left(\frac{\partial^2 h}{\partial x^2} \right) (\delta x)^2 = \frac{1}{2}\mathbf{D}_x^2(h)(\delta x)^2 \tag{2.4.6}$$

Now, substituting (2.4.1)–(2.4.3) into (2.4.5) we get

$$\Delta \mathbf{h} = \mathbf{D}_x(h) \left[\mathbf{D}_c(x)\zeta + \frac{1}{2}\zeta^T \mathbf{D}_c^2(x)\zeta \right]$$

$$+ \frac{1}{2}\mathbf{D}_x^2(h) \left[\mathbf{D}_c(x)\zeta + \frac{1}{2}\zeta^T \mathbf{D}_c^2(x)\zeta \right]^2 \tag{2.4.7}$$

which is already a 4th-degree polynomial in the components of ζ.

Since we are interested in the second-order approximation, dropping the terms of degree 3 or more in $\boldsymbol{\zeta}$ from (2.4.7), we obtain[1]

$$\Delta\mathbf{h} = \mathbf{H}\boldsymbol{\zeta} + \boldsymbol{\zeta}^T A \boldsymbol{\zeta} \tag{2.4.8}$$

where

$$\mathbf{H} = \mathbf{D}_x(h)\mathbf{D}_c(x) \in \mathbb{R}^{1 \times 2}$$
$$A = \mathbf{a}_1 \mathbf{D}_c^2(x) + \mathbf{a}_2\, \mathbf{D}_c^T(x)\, \mathbf{D}_c(x) \in \mathbb{R}^{2 \times 2} \tag{2.4.9}$$
$$\mathbf{a}_1 = \frac{1}{2}D_x(h) \quad\text{and}\quad \mathbf{a}_2 = \frac{1}{2}\mathbf{D}_x^2(h).$$

If $h(x) = x$, then $\mathbf{D}_x(h) = 1$ and $\mathbf{D}_x^2(h) = 0$. In this case, $a_1 = 1/2$ and $a_2 = 0$, and A is the Hessian multiplied by 0.5. Our goal is to find $\boldsymbol{\zeta}$ such that

$$\mathbf{z}(k) \approx h(x(k) + \Delta x) = \mathbf{h}(x(k)) + \Delta h \tag{2.4.10}$$

or

$$\Delta\mathbf{h} = \mathbf{e}_F(k) = \mathbf{z}(k) - \mathbf{h}(x(k)).$$

Substituting for Δh from (2.4.8), we obtain the following constraint associated with the FSM process.

$$\mathbf{g}(\boldsymbol{\zeta}) = \mathbf{H}\boldsymbol{\zeta} + \boldsymbol{\zeta}^T A \boldsymbol{\zeta} = \mathbf{e}_F. \tag{2.4.11}$$

We do not demand that this constraint be satisfied exactly, but that the squares of the residual associated with this equation be minimized. This condition is expressed as minimization of

$$\mathbf{f}(\boldsymbol{\zeta}) = \frac{1}{2}\left[\mathbf{e}_F - \mathbf{H}\boldsymbol{\zeta} - \boldsymbol{\zeta}^T A \boldsymbol{\zeta}\right]^2, \tag{2.4.12}$$

which is a polynomial of degree 4 in the components of $\boldsymbol{\zeta}$ (Lakshmivarahan et al. 2003). We drop all terms of degree 3 or more from the right hand side of (2.4.12), and obtain the full quadratic approximation $\mathbf{Q}(\boldsymbol{\zeta})$ to $\mathbf{f}(\boldsymbol{\zeta})$ given by

$$\mathbf{Q}(\boldsymbol{\zeta}) = \frac{1}{2}\left[\mathbf{e}_F - 2\mathbf{e}_F\left(\mathbf{H}\boldsymbol{\zeta} + \boldsymbol{\zeta}^T A \boldsymbol{\zeta}\right) + \boldsymbol{\zeta}^T \mathbf{H}^T \mathbf{H}\boldsymbol{\zeta}\right]$$
$$= \frac{1}{2}\left[\mathbf{e}_F^2 - 2\mathbf{e}_F\mathbf{H}\boldsymbol{\zeta} + \boldsymbol{\zeta}^T\left(\mathbf{H}^T \mathbf{H} - 2\mathbf{e}_F A\right)\boldsymbol{\zeta}\right]. \tag{2.4.13}$$

---

[1]Using $(\mathbf{a}^T\mathbf{x})^2 = \mathbf{a}^T\mathbf{x}\,\mathbf{a}^T\mathbf{x} = \mathbf{x}^T(\mathbf{a}\mathbf{a}^T)\mathbf{x}$, where $\mathbf{a}$ and $\mathbf{x}$ are column vectors.

Setting the gradient of $Q(\zeta)$ with respect to ζ equal to zero, we obtain

$$\left[\mathbf{H}^T\mathbf{H} - 2e_F\mathbf{A}\right]\zeta = \mathbf{H}^T\mathbf{e}_F \tag{2.4.14}$$

The solution of (2.4.14) is indeed a minimum provided, the Hessian of $\mathbf{Q}(\zeta)$, using (2.4.9), is given by

$$\nabla^2\mathbf{Q}(\zeta) = \mathbf{H}^T\mathbf{H} - 2e_F\mathbf{A}$$

$$= \mathbf{H}^T\mathbf{H} - \mathbf{e}_F\left[\mathbf{D}_x(h)\mathbf{D}_\zeta^2(x) + \mathbf{D}_x^2(h)\mathbf{D}_\zeta^T(x)\mathbf{D}_\zeta(x)\right] \tag{2.4.15}$$

is positive definite.

When two or more observations are considered, the stacked form of the sensitivities and error as found in (2.2.25) come into play. In place of (2.4.14), we get

$$\left[\mathbf{H}^T\mathbf{H} - 2e_F(t_1)A(t_1) - e_F(t_2)A(t_2)\cdots - e_F(t_N)A(t_N)\right]\zeta = \mathbf{H}^T\mathbf{e}_F \tag{2.4.16}$$

2.5 FSM: Discrete Time Formulation

In Sects. 2.1 and 2.2 we have illustrated the basic principles that underlie the first-order forward sensitivity method (FSM) using a simple, scalar model in continuous time. However, in practice without exception, all the large scale models operate in discrete time. As a prelude to this transition, we derive the dynamics of first-order and second-order sensitivity evolution in discrete time.

2.5.1 Discrete Evolution of First-Order Forward Sensitivities

Discretizing equation (2.1.1) using the standard Euler scheme (Richtmyer and Morton 1967), we obtain

$$x(k+1) = \mathbf{M}(x(k),\alpha). \tag{2.5.1}$$

where $M : \mathbb{R} \times \mathbb{R} \to \mathbb{R}$ is a nonlinear function given by

$$M(x(k),\alpha) = x(k) + \Delta t f(x(k),\alpha) \tag{2.5.2}$$

where Δt is the length interval used in time discretization. Differentiating (2.5.1) with respect to $x(0)$, we readily obtain the dynamics of evolution of the first-order forward sensitivity $\partial x(k)/\partial x(0)$ as

$$\frac{\partial x(k+1)}{\partial x(0)} = \frac{\partial M(x(k),\alpha)}{\partial x(k)}\frac{\partial x(k)}{\partial x(0)} \tag{2.5.3}$$

with

$$\frac{\partial x(k)}{\partial x(0)}\bigg|_{k=0} = 1$$

as the initial condition. Much like its continuous time counterpart, (2.5.3) is a linear, homogeneous and non-autonomous first-order difference equation. Setting

$$u_1(k) = \frac{\partial x(k)}{\partial x(0)}, \quad \text{and} \quad \mathbf{D}_x(M) = \frac{\partial M(x(k), \alpha)}{\partial x(k)}, \tag{2.5.4}$$

the discrete time evolution in (2.5.3) can be written as

$$u_1(k+1) = \mathbf{D}_{x(k)}(\mathbf{M}) \, u_1(k) \quad \text{with} \quad u_1(0) = 1. \tag{2.5.5}$$

Alternatively, one could also obtain (2.5.3) by discretizing (2.1.3) using the Euler scheme (Problem 2.1)

$$\mathbf{D}_x(M) = 1 + \Delta t \mathbf{D}_x(f). \tag{2.5.6}$$

Similarly, differentiating (2.5.1) with respect to α, we get

$$\frac{\partial x(k+1)}{\partial \alpha} = \frac{\partial M(x(k), \alpha)}{\partial x(k)} \frac{\partial x(k)}{\partial \alpha} + \frac{\partial M(x(k), \alpha)}{\partial \alpha} \tag{2.5.7}$$

Setting

$$v_1(k) = \frac{\partial x(k)}{\partial \alpha} \quad \text{and} \quad \mathbf{D}_\alpha = \frac{\partial M(x(k), \alpha)}{\partial \alpha} \tag{2.5.8}$$

in (2.5.7), the latter becomes

$$v_1(k+1) = \mathbf{D}_{x(k)}(M) v_1(k) + \mathbf{D}_\alpha(M) \tag{2.5.9}$$

which is the discrete time analog of (2.1.8) for the evolution of the forward sensitivity $v_1(k)$. Derivation of (2.5.9) by directly discretizing (2.1.8) is left as the Problem 2.1.

Example 2. Consider the discrete time version of the logistic model for population dynamics given by

$$x(k+1) = \alpha x(k) (1 - x(k)) \quad \text{for } \alpha > 0. \tag{2.5.10}$$

Then $M(x, \alpha) = \alpha x(1 - x)$ and $\mathbf{D}_x(M) = \alpha(1 - 2x)$, and $\mathbf{D}_\alpha(M) = x(1 - x)$.

Hence

$$\frac{\partial x(k+1)}{\partial x(0)} = \alpha \, (1 - 2x(k)) \, \frac{\partial x(k)}{\partial x(0)} \quad \text{with} \quad \frac{\partial x(0)}{\partial x(0)} = 1, \qquad (2.5.11)$$

and

$$\frac{\partial x(k+1)}{\partial \alpha} = \alpha \, (1 - 2x(k)) \, \frac{\partial x(k)}{\partial \alpha} + x(k)(1 - x(k)), \text{ with}$$

$$\frac{\partial x(0)}{\partial \alpha} = 0. \qquad (2.5.12)$$

Plots of the solution of (2.5.11) and (2.5.12) are given in Figs. 2.6 and 2.7. Figure 2.8 contains the plot of the asymptotic state $x(0)$ for various values of $1 \le \alpha \le 4$. Figure 2.8 is obtained by fixing a value of the parameter in the range $1 \le \alpha \le 4$ in steps of 0.001 and running the model 2.5.10 for a long time until steady state $x(\infty)$ is reached. For $1 \le \alpha \le 3$, there is a unique steady state. For $3 < \alpha \le 3.45$, we observe the first bifurcation, called periodic doubling where $x(\infty)$ oscillates between two values. Further periodic doubling occurs where $3.45 \le \alpha \le 3.54$ and the system exhibits non-chaotic behavior for $\alpha < 3.57$. When α is around 3.57, there is a cascading of periodic doubling behavior which signals the onset of chaos. An expanded view of Fig. 2.8 for $3 \le \alpha \le 4$ is given in Fig. 2.9

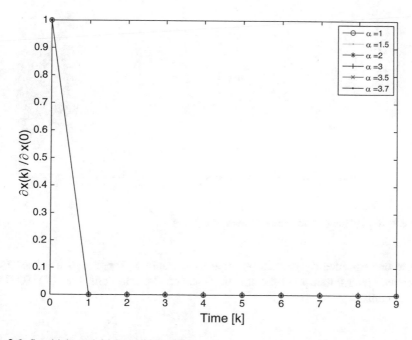

Fig. 2.6 Sensitivity to initial condition $x(0)$

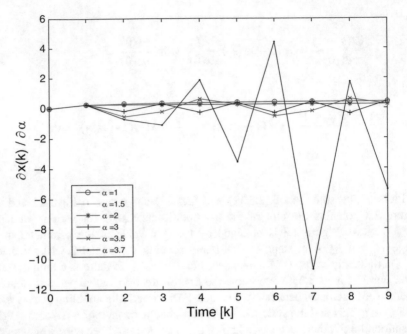

Fig. 2.7 Sensitivity to value of α

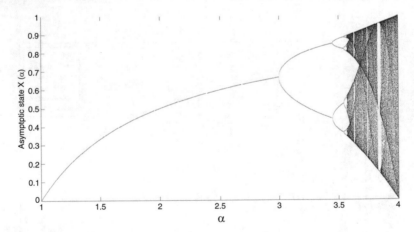

Fig. 2.8 Asymptotic state for various values of $1 \leq \alpha \leq 4$

where one can identify gaps in the final state diagram. These gaps are due to the so called "intermittency" of the chaotic behavior. Refer to Peitgen et al. (1992) for more details.

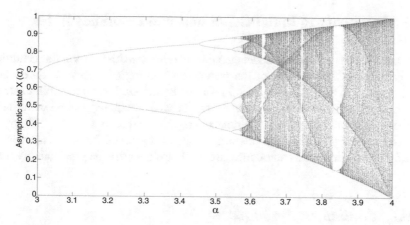

Fig. 2.9 Asymptotic state for various values of $3 \le \alpha \le 4$

2.5.2 Discrete Evolution of Second-Order Forward Sensitivity

By following the first principles used in the derivation of dynamics of evolution of second-order forward sensitivities in Sect. 2.5, it can be verified that the discrete time evolution of second-order sensitivities are given by (Problem 2.2).

$$\frac{\partial^2 x(k+1)}{\partial x^2(0)} = \frac{\partial M}{\partial x(k)} \frac{\partial^2 x(k)}{\partial x^2(0)} + \frac{\partial^2 M}{\partial x^2(k)} \left(\frac{\partial x(k)}{\partial x(0)} \right)^2 \tag{2.5.13}$$

$$\frac{\partial^2 x(k+1)}{\partial \alpha^2} = \frac{\partial M}{\partial x(k)} \frac{\partial^2 x(k)}{\partial \alpha^2} + \frac{\partial^2 M}{\partial x^2(k)} \left(\frac{\partial x(k)}{\partial \alpha} \right)^2$$
$$+ 2 \left(\frac{\partial^2 M}{\partial x(0) \partial \alpha} \right) \left(\frac{\partial x(k)}{\partial \alpha} \right) + \frac{\partial^2 M}{\partial \alpha^2} \tag{2.5.14}$$

and

$$\frac{\partial^2 x(k+1)}{\partial x(0) \partial \alpha} = \left(\frac{\partial M}{\partial x(k)} \right) \left(\frac{\partial^2 x(k)}{\partial x(0) \partial \alpha} \right)$$
$$+ \frac{\partial^2 M}{\partial x^2(k)} \frac{\partial x(k)}{\partial \alpha} \frac{\partial x(k)}{\partial x(0)} + \left(\frac{\partial^2 M}{\partial x(k) \partial \alpha} \right) \frac{\partial x(k)}{\partial x(0)} \tag{2.5.15}$$

2.6   Sensitivity to Initial Conditions and Lyapunov Index

As stated in Chap. 1, some models exhibit extreme sensitivity to initial conditions for certain ranges of values of the parameter. The logistic model is an example we discussed in Sect. 2.5 (Example 2)—a text book case known to exhibit extreme sensitivity to initial conditions when $\alpha = 4$. In such situations, a slight perturbation to the initial condition grows with time and masks the quality forecast. This growth of error naturally leads to predictability limits as discussed further in this section. The Lyapunov index is a natural measure that helps quantify the predictability limit.

2.6.1   Continuous Time Model

Consider the forecast model of the type (2.1.1). Let $\bar{x}(t)$ be the base trajectory of this model starting from the base initial condition $\bar{x}(0)$. Let

$$x(0) = \bar{x}(0) + y(0) \qquad (2.6.1)$$

be the modified initial condition obtained by adding a perturbation $y(0)$ to $\bar{x}(0)$. Let $x(t)$ be the solution of (2.1.1) starting from $x(0)$. Then

$$y(t) = x(t) - \bar{x}(t). \qquad (2.6.2)$$

Differentiating both sides of (2.6.2) and using (2.1.1), we get

$$\dot{y}(t) = f(x(t), \alpha) - f(\bar{x}(t), \alpha). \qquad (2.6.3)$$

Substituting (2.6.2) into the first term on the right-hand side of (2.6.3) and expanding in the first-order Taylor series, we readily obtain the dynamics of perturbation evolution:

$$\dot{y}(t) = \mathbf{D}_{\bar{x}(t)}(f)y(t). \qquad (2.6.4)$$

This is a scalar linear, non-autonomous O.D.E. of the same type as the FSM dynamics in (2.1.5). Equation (2.6.4) is also called the *tangent linear system* or the *variational equation* in the literature. The initial condition for (2.1.5) is $u(0) = \left.\frac{\partial x(t)}{\partial x(0)}\right|_{t=0} = 1$, while that of (2.6.4) is $y(0)$, which is the perturbation on the initial condition $\bar{x}(0)$.

   Let $[0, T]$ be the time interval that denotes the forecast horizon. Discretize this time interval into N subintervals of equal length τ, where $T = N\tau$. It is reasonable to assume that the Jacobian $\mathbf{D}_{\bar{x}(t)}(f)$ remains nearly constant in the kth subinterval $[(k-1)\tau, k\tau]$ by choosing an appropriate value of τ. Define

$$L_{k-1} = \left.\mathbf{D}_{\bar{x}(t)}(f)\right|_{t=(k-1)\tau} \qquad (2.6.5)$$

Then, on this kth subinterval $(k-1)\tau \le t \le k\tau$, we can represent (2.6.4) as

$$\dot{y} = L_{k-1}y \qquad (2.6.6)$$

for $k = 1, 2, \ldots, N$. Solving (2.6.6), we get

$$y_k = y_{k-1}e^{L_{k-1}\tau}, \qquad (2.6.7)$$

where $y_k = y(t = k\tau)$. Clearly $e^{L_{k-1}\tau}$ is the error magnification during the kth subinterval. Iterating (2.6.7), it follows that

$$y_k = y_0 e^{\left(\sum_{k=0}^{N-1} L_k\right)\tau} \qquad (2.6.8)$$

Thus, the perturbation y_k at time k is related to that at time $k = 0$ through a magnification/amplification factor that is given by the exponential term on the right hand side of (2.6.8). The Lyapunov index λ is defined by

$$\lambda = \lim_{T \to \infty} \lim_{N \to \infty} \frac{1}{(N\tau)} \ln \left(\frac{|y_k|}{|y_0|} \right) \qquad (2.6.9)$$

Thus, when T is large and τ is very small, at time $t = k\tau$, we can express

$$y_k \approx e^{\lambda t} y_0 \qquad (2.6.10)$$

Using (2.6.8) in (2.6.9), it readily follows that

$$\lambda = \lim_{T \to \infty} \lim_{N \to \infty} \frac{1}{N} \left(\sum_{k=0}^{N-1} L_k \right). \qquad (2.6.11)$$

Thus, λ denotes the long term average growth rate of errors.

2.6.2   Discrete Time Model

Starting from the discrete time model in (2.5.1), let $\bar{x}(k)$ be the base trajectory starting from $\bar{x}(0)$. If $x(0) = \bar{x}(0) + y(0)$ with $y(0)$ as the initial perturbation, then the perturbation $y(k) = x(k) - \bar{x}(k)$ is given by

$$\mathbf{y}(k) = \mathbf{M}(x(k, \alpha)) - \mathbf{M}(\bar{x}(k), \alpha). \qquad (2.6.12)$$

Expanding the first term on the right hand side in a first-order Taylor series around $\bar{x}(k)$, the dynamics of evolution of the perturbation is given by the tangent linear system

$$\mathbf{y}(k+1) = \mathbf{D}_{\bar{x}(k)}(\mathbf{M})y(k) \qquad (2.6.13)$$

Iterating, we obtain (refer to Problem 2.5)

$$\mathbf{y}(k) = \left[\prod_{i=0}^{k-1}{}' A(i) \right] y(0). \tag{2.6.14}$$

where $A(i) = \mathbf{D}_{\bar{x}(i)}(\mathbf{M})$ for simplicity in notation. From Problem 2.5, it can be verified that

$$y(k) = u_1(k)y(0), \tag{2.6.15}$$

where

$$u_1(k) = \prod_{i=0}^{k-1}{}' \mathbf{D}_{x(i)}(M) = \frac{\partial \bar{x}(k)}{\partial \bar{x}(0)}, \tag{2.6.16}$$

the forward sensitivity of the solution $\bar{x}(k)$ with respect to $\bar{x}(0)$. Define

$$\Lambda(k) = \left(\frac{y(k)}{y(0)} \right)^{1/k} = \left[\prod_{i=0}^{k-1} A(i) \right]^{1/k} = [u_1(k)]^{1/k} \tag{2.6.17}$$

which is the geometric mean of the forward sensitivity at time k. Then, from the definition, the expression for the Lyapunov index is given by

$$\lambda = \lim_{k \to \infty} \ln(\Lambda(k))$$

$$= \lim_{k \to \infty} \frac{1}{k} \sum_{i=0}^{k-1} \ln A(i)$$

$$= \lim_{k \to \infty} \frac{1}{k} \ln u_1(k), \tag{2.6.18}$$

which is the long-term average of the logarithm of the first-order forward sensitivities.

2.7 Exercises

Problem 2.1. By directly discretizing (2.1.3) and (2.1.8) using the standard Euler scheme, obtain the corresponding discrete dynamics given in (2.5.3) and (2.5.9) respectively.

Problem 2.2. Starting from (2.5.3) and (2.5.9), derive the dynamics of the evolution of second-order forward sensitivities given in (2.5.13)–(2.5.15) .

Problem 2.3. Consider a linear, first-order, time varying and non-homogeneous recurrence relation

$$\mathbf{y}(k+1) = \mathbf{A}(k)\mathbf{y}(k) + \mathbf{B}(k),$$

where $\mathbf{y}(k) \in \mathbb{R}^n, \mathbf{A}(k) \in \mathbb{R}^{n \times n}$ and $\mathbf{B}(k) \in \mathbb{R}^n$ with $\mathbf{y}(0) \in \mathbb{R}^n$ as the initial condition. We illustrate the *method of substitution* for solving the above recurrence relation. Clearly

$$\mathbf{y}(1) = \mathbf{A}(0)\mathbf{y}(0) + \mathbf{B}(0),$$
$$\mathbf{y}(2) = \mathbf{A}(1)y(1) + \mathbf{B}(1)$$
$$= \mathbf{A}(1)\mathbf{A}(0)y(0) + \mathbf{A}(1)B(0) + \mathbf{B}(1),$$
$$\mathbf{y}(3) = \mathbf{A}(2)\mathbf{y}(2) + \mathbf{B}(2)$$
$$= \mathbf{A}(2)\mathbf{A}(1)\mathbf{A}(0)\mathbf{y}(0) + \mathbf{A}(2)\mathbf{A}(1)\mathbf{B}(0) + \mathbf{A}(2)\mathbf{B}(1) + \mathbf{B}(1).$$

Verify inductively that

$$\mathbf{y}(k) = \left[\prod_{i=0}^{k-1}{'} \mathbf{A}(i) \right] \mathbf{y}(0) + \sum_{i=0}^{k-1} \left[\prod_{q=i+1}^{k-1} \mathbf{A}(q) \right] \mathbf{B}(i)$$

where to indicate the order of indices in the product we use two symbols:

$$\prod_{i=0}^{k} \mathbf{A}(i) = \mathbf{A}(0)\mathbf{A}(1)\ldots\mathbf{A}(k)\text{---natural order}$$

$$\prod_{i=0}^{k}{'} \mathbf{A}(i) = \mathbf{A}(k)\mathbf{A}(k-1)\ldots\mathbf{A}(0)\text{---reverse order}$$

and

$$\prod_{i=p}^{q} \mathbf{A}(i) = 1 \text{ if } q < p\text{---vacuous product}.$$

While these two products are the same when $\mathbf{A}(i)$ are scalars, they are distinct when $\mathbf{A}(i)$ are matrices.

Problem 2.4. Consider the linear, first-order, time varying and homogeneous recurrence (2.5.9) that describes the evolution of the first-order sensitivity vector $\mathbf{v}_1(k) \in \mathbb{R}^p$, of $x(k)$ with respect to $\boldsymbol{\alpha} \in \mathbb{R}^p$ given by

$$\mathbf{v}_1(k+1) = \mathbf{A}(k)\mathbf{v}_1(k) + \mathbf{B}(k), \text{ with } \mathbf{v}_1(0) = 0 \in \mathbb{R}^p$$

where $\mathbf{A}(k) = \mathbf{D}_{x(k)}(\mathbf{M})$ and $\mathbf{B}(k) = \mathbf{D}_\alpha(\mathbf{M}(x(k), \alpha))$ for simplicity in notation. The recurrence is structurally the same as the one in Exercise 2.3 given above. By setting $\mathbf{y}(k) = \mathbf{v}_1(k)$ and knowing that $\mathbf{v}_1(0) = 0$, verify that the solution $\mathbf{v}_1(k)$ is given by

$$\mathbf{D}_\alpha\left(x(k)\right) = \frac{\partial x(k)}{\partial \alpha} = \mathbf{v}_1(k) = \sum_{i=0}^{k-1}\left[\prod_{q=i+1}^{k-1}\mathbf{A}(q)\right]\mathbf{B}(i).$$

Problem 2.5. Consider the scalar, linear, time varying and homogeneous recurrence relation (2.5.5) that describes the evolution of the first-order sensitivity, $u_1(k) \in \mathbb{R}$ of the solution $x(k)$ with respect to the initial condition $x(0)$. It is given by

$$u_1(k + 1) = A(k)u_1(k)$$

with $A(k) = \mathbf{D}_{x(k)}(\mathbf{M})$ and $u_1(0) = 1$. By setting $n = 1, y(k) = u_1(k)$ and $B(k) \equiv 0$ in the solution of the recurrence in Problem 2.3, verify that (since $u_1(0) = 1$) that

$$\mathbf{D}_{x(0)}\left(x(k)\right) = \frac{\partial x(k)}{\partial x(0)} = u_1(k) = \prod_{i=0}^{k-1}A(i),$$

which is the product of the sensitivities of $x(k)$ along the trajectory.

Problem 2.6. Consider the logistic equation in continuous-time form with carrying capacity 1 $(0 \leq x \leq 1)$ and growth parameter α given by

$$\frac{\partial x}{\partial t} = \alpha x\,(1 - x) \quad \text{with} \quad x(0) = x_0.$$

a) Verify that the solution is given by

$$x(t) = \frac{x_0 e^{\alpha t}}{1 - x_0 + x_0^{\alpha t}}.$$

   Plot $x(t)$ versus t when $x_0 = 0.5$ and $\alpha = 1.0$.
b) Compute $\partial x(t)/\partial x_0$ and $\partial x(t)/\partial \alpha$ and plot their evolutions.
c) Compute $\partial^2 x(t)/\partial x_0^2$, $\partial^2 x(t)/\partial \alpha^2$ and $\partial^2 x(t)/\partial x_0 \partial \alpha$ and also plot their evolutions.

Problem 2.7. [From Rabitz et al. (1983)] Consider a partial differential equation of type

$$\frac{\partial u}{\partial t} = D\frac{\partial^2 u}{\partial x^2} + f(u, \alpha). \quad (*)$$

where D is a diffusion constant. When $u = u(x, 0)$, $t \geq 0$, $x_1 \leq x \leq x_2, f(u, \alpha)$ is some (nonlinear) function of u and the scalar parameter α. The initial condition for $u(x, t)$ is given by $u(x, 0) = g(x)$ for all $x_1 \leq x \leq x_2$ and the boundary condition for $u(x, t)$ is given by

$$a_1 \frac{\partial u}{\partial x}\bigg|_{x=x_1} + b_1 u(x, t)\bigg|_{x=x_1} = A_1,$$

$$a_2 \frac{\partial u}{\partial x}\bigg|_{x=x_2} + b_2 u(x, t)\bigg|_{x=x_2} = A_2$$

where a_i, b_i, A_i $(1 \leq i \leq 2)$ are constants. Let $s(x, t) = \partial u(x, t)/\partial \alpha$ denote the sensitivity of the solution $u(x, t)$ with respect to α. Differentiating both sides of the partial differential equation $(*)$ to verify that $s(x, t)$ is given by the solution of the following system

$$\frac{\partial s}{\partial t} = D \frac{\partial^2 s}{\partial x^2} + \frac{\partial f}{\partial u} s + \frac{\partial f}{\partial \alpha},$$

where $s(x, 0) = \partial g/\partial \alpha$ is the initial condition and the boundary conditions are given by

$$a_1 \frac{\partial s}{\partial \alpha}\bigg|_{x=x_1} + b_1 s\bigg|_{x=x_1} = 0,$$

$$a_2 \frac{\partial s}{\partial \alpha}\bigg|_{x=x_2} + b_2 s\bigg|_{x=x_2} = 0.$$

Problem 2.8. Using (2.6.18) compute and plot the variation of the Lyapunov index with respect to the parameter α for the discrete time logistic model in Example 2 of Sect. 2.4.

2.7.1 Demonstrations

Demonstration: Air–Sea Interaction Model

Using the development in Example 1, Sect. 2.1, we solve the FSM data assimilation problem based on the dynamics of air–sea interaction. The solution take the analytic form

$$x(t, x_0, \theta, c) = (x(0) - \theta)e^{-ct} + \theta$$

as previously shown in (2.1.12). The development of formulas for first
and second-order sensitivities, applicable to the air–sea interaction dynamics, appears in Sect. 2.1 and 2.3, respectively. The form of the FSM data
assimilation process has been developed in Sect. 2.2 (first-order) and 2.4
(second-order).

In this demonstration, we test the FSM using: (i) first-order sensitivity (1
iteration), (ii) first-order sensitivity (2 iterations), (iii) second-order sensitivity
dropping terms of power 3 and 4 in the functional [labeled second-order I],
and (iv) second-order sensitivity including all power of control (up to power 4)
[labeled second-order II]. The true control $c\,(x, \theta, c)$ and erroneous control c'
are taken to be:

$$c = (1.0, 11.0, 0.25)$$

$$c' = (3.0, 9.0, 0.35)$$

Forecast is made from erroneous control in the presence of four observations. The observations, forecast, and increment (observation minus forecast)
are displayed in Table 2.1.

Table 2.1 Observation (obs, x), forecast (fcst, x) and increment (obs–fcst) for numerical experiments

obs index	Time (h)	obs	fcst	Increment
1	1.0	3.49	4.77	−1.27
2	2.0	5.31	6.02	−0.71
3	8.0	8.93	8.64	0.29
4	12.0	10.28	8.91	1.37

Display of the forecast and observations is shown in Fig. ASI.1 where
the Guess temperature is the forecast from erroneous control. The optimally
adjusted forecasts from the four experiments are displayed in the two panels of
Fig. ASI.2 (first-order FSM) and Fig. ASI.3 (second-order FSM). The optimal
adjusted control for each case is shown in Table 2.2.

Table 2.2 Control for the various experiments

Control	X_0	θ	c
True	1.0	11.0	0.25
Erroneous	3.0	9.0	0.35
First-order (1 iter)	1.89	10.32	0.17
First-order (2 iter)	2.14	10.90	0.20
Second-order I	1.61	10.64	0.14
Second-order II	2.0	10.52	0.22

Results indicate that the 2-step first-order sensitivity method gives an excellent result. The 1-step first order method is acceptable. The second-order I is biased low. The second-order II is nearly perfect.

It may at first seem puzzling that the second-order method (full quadratic form) gave such a poor result. But the general philosophy of second order methods is that they provide value if the operating point is "close" to the local minimum—near the "well". Thus, FSM first-order method can be used to move close to the local minimum, and then use the second-order (full quadratic) to get closer to the minimum. We leave this as an exercise for the student. For most problems, the second-order method using all terms up to fourth order is extremely difficult or impossible to solve. In this case, steepest descent algorithm was used.

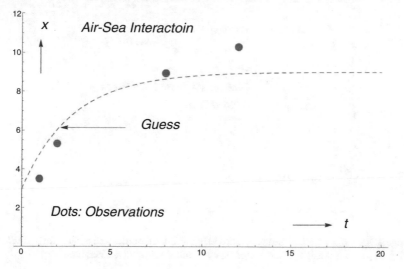

Fig. ASI.1 Guess temperature $[x(t)]$ evolution with observations superimposed—sea–air interaction

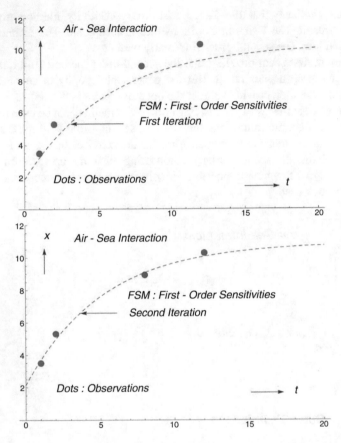

Fig. ASI.2 *Top panel*: Adjusted profile using FSM first-order sensitivity, first iteration. *Bottom panel*: Adjusted profile using FSM first-order sensitivity, 2 iterations

Demonstration: Gauss' Problem

In Lewis et al. (2006), a detailed account of Carl Friedrich Gauss' discovery of the method of least squares under dynamical constraint has been presented. The method was developed when Gauss succeeded in forecasting the time and place of reappearance of the "unknown planet" after its conjunction with the sun. There was only a limited number of observations of this celestial object between January 1, 1801, and February 11, 1801—a 42-day period of observation. The object was actually a large asteroid later named Ceres. Gauss' theoretical development of methodology that led to accurate prediction

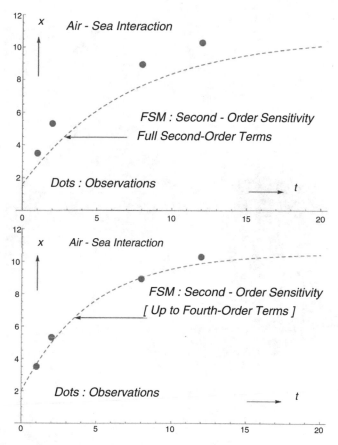

Fig. ASI.3 Adjusted profile using FSM second-order sensitivity (full quadratic form). *Bottom panel*: Adjusted profile using FSM second-order sensitivity (including terms up to fourth order in the functional)

of Ceres' orbit was published in November of the same year. Gauss' results were used to locate the asteroid on January 1, 1802, exactly 1 year after its first sighting. Figure GS.1 is a drawing of Gauss where the background schematic depicts three observations of Ceres' location relative to earth before it comes into conjunction with the sun.

In Lewis et al. (2006), a simplified version of Gauss' method of solution was discussed in detail with an outline of solution methodology based on 4D-VAR. The simplification was essentially the assumption that the asteroid and the earth moved about the sun along circular trajectories rather than elliptical trajectories.

Fig. GS.1 Carl Friedrich Gauss in his academic garb shown aside a schematic showing positions of Ceres as observed from Earth in early 1801 (*small circled dots*) and in early 1802 (*small filled circle*)

We now approach the problem using FSM. The geometry of the problem is shown in Figs. GS.2 and GS.3, exactly the same figures used in Lewis et al. (2006)—Figs. 4.5.1 and 4.5.2 in that book. The nomenclature is described on the insert of Fig. GS.2 where we have assumed that the three bodies are initially co-planar and remain so into the future.

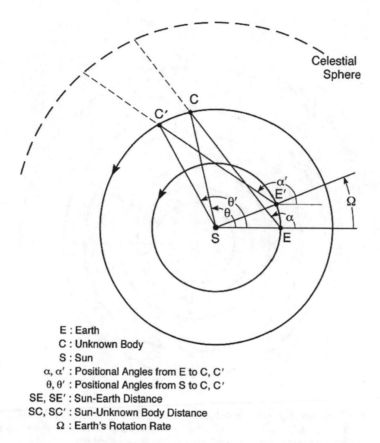

E : Earth
C : Unknown Body
S : Sun
α, α' : Positional Angles from E to C, C'
θ, θ' : Positional Angles from S to C, C'
SE, SE' : Sun-Earth Distance
SC, SC' : Sun-Unknown Body Distance
Ω : Earth's Rotation Rate

Fig. GS.2 Geometric depiction of the orbits of Earth (E) and Ceres (C) around the Sun (S) with angular measurements from Earth to Ceres against the celestial sphere backdrop

Kepler's third law is the dynamical constraint. Namely, each body in revolution about the sun exhibits constancy of the ratio $\frac{r^3}{T^2}$ where r is the distance from the sun and T is the period of rotation about the sun. We use the orbital elements of earth to evaluate the constant—$r = 1$ A.U. (astronomical unit) and $T = 1$ year. Thus, the constant is 1 (A.U.)^3(year)$^{-2}$.

The asteroid Ceres lies along the line with angle α relative to SE at $t = 0$. At a later time Δt, Fig. GS.2 shows that Ceres lies along the line with angle $\alpha\prime$ relative to SE (earth has rotated through an angle $= \Omega \Delta t$, where $\Omega = 2\pi$ radians/year). The problem reduces to finding Ceres' distance from the sun, its period of revolution about the sun, and its initial angle Θ relative to SE. This is the control vector for the problem. To find these elements of control, at least three observations of the angles $\alpha, \alpha', \ldots$ are required. But since these observations typically include noise (error), the problem is best solved in a least squares context with more than three observations.

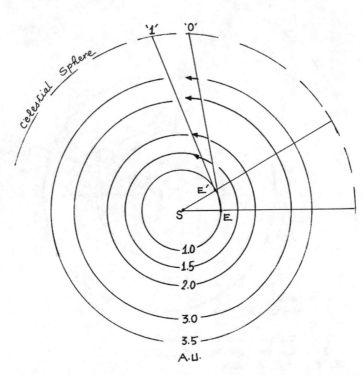

Fig. GS.3 Motion of objects at various distances from the Sun [Earth (E) at 1.0 A.U., other objects at 1.5, 2.0, 3.0, and 3.5 A.U.'s] and their 1-month angular movement according to Kepler's third Law after starting at locations along the line-of-sight EO

Before discussing the mathematical steps involved in solving the problem, it is instructive to qualitatively view the problem for a geometric perspective. Figure GS.3 depicts the case where Ceres is first sighted along the line from E to the point "O" on the celestial sphere—the background of fixed stars. After 1 month, the earth moves 30° along its circular trajectory and is located at E'. As stated earlier, the radius of Earth's trajectory relative to the sun's fixed location is 1 A.U. At $t = 1$ month, we find Ceres along the line from E' to point "1" on the celestial sphere. Now, using Kepler's third law, we move the points initially along EO to points along their respective trajectories 1 month later. These circular displacements are shown by curved arrows from the initial location (tail of arrow) to the location after 1 month (head of arrow). The displacement for the point on the 2 A.U. circle exactly falls on the E'1 line, the other displacements are either too large (1.5 A.U. circle) or too small (3.0 and 3.5 A.U. circles). Thus, in this idealized example, the asteroid is located at 2.0 A.U.'s from the sun. It thus becomes clear that measurement of the "α" angles as a function of time hold the power to determine the control elements.

Now for some mathematical details. The simplest approach to developing formulas for this problem is to make use of theory of complex variables. Let us represent the radius of earth's orbit by $R = 1$ A.U. and the unknown radius of Ceres' orbit by r. The revolution periods of these bodies about the Sun are represented by Ω (Earth) and $d\Theta/dt(= \dot{\Theta})$ for Ceres. We measure the angles α and Θ relative to SE (the line from the sun to earth at $t = 0$). Thus the positions of the Earth and Ceres at various times t are

$$R e^{i\Omega t} \quad \text{(earth)}$$
$$r e^{i(\Theta + \dot{\Theta}t)} \quad \text{(Ceres)} . \tag{GS.1}$$

The line in the complex plane connecting Earth and Ceres (line from earth to Ceres) is

$$D(r, \Theta, \dot{\Theta}) = r e^{i(\Theta + \dot{\Theta}t)} - R e^{i\Omega t} \tag{GS.2}$$

where R and Ω are known. This complex number D can be decomposed into its real and imaginary parts and from these components we can find the angle α as a function of time from Earth to Ceres. Accordingly,

$$D = \begin{cases} r\cos(\Theta + \dot{\Theta}t) - R\cos(\Omega t) \\ +i[r\sin(\Theta + \dot{\Theta}t) - R\sin(\Omega t)] \end{cases} \tag{GS.3}$$

$$\alpha = \arctan \frac{r\sin(\Theta + \dot{\Theta}t) - R\sin(\Omega t)}{r\cos(\Theta + \dot{\Theta}t) - R\cos(\Omega t)}$$

Let

$$x = r\cos(\Theta + \dot{\Theta}t) - R\cos(\Omega t)$$
$$y = r\sin(\Theta + \dot{\Theta}t) - R\sin(\Omega t). \tag{GS.4}$$

Then

$$\alpha = \arctan\left(\frac{y}{x}\right). \tag{GS.5}$$

Taking the derivative of α with respect to time, we get

$$\dot{\alpha} = \frac{\dot{y} - \dot{x}\tan\alpha}{x\sec^2\alpha} \tag{GS.6}$$

This is the forecast equation for the problem. In the FSM approach, we need to find sensitivities of α with respect to the elements of control r, $\dot{\Theta}$, and Θ.

We then minimize the difference between the measured values of α and the adjusted forecast—the guess forecast (using control guess) added to first-order Taylor series terms $\frac{\partial \alpha}{\partial r} \Delta r$, $\frac{\partial \alpha}{\partial \dot{\Theta}} \Delta \dot{\Theta}$, and $\frac{\partial \alpha}{\partial \Theta} \Delta \Theta$.

2.8   Notes and References

This chapter follows the development in Lakshmivarahan and Lewis (2010). The classic review by Rabitz et al. (1983) provides a comprehensive account of the theory of parameter sensitivity and its varied applications to chemical kinetics. A complete and thorough discussion of the concept of Lyapunov index and its computation is given in Peitgen et al. (1992).

Chapter 3
On the Relation Between Adjoint and Forward Sensitivity

Our goal in this chapter is to clarify the intrinsic relation that exists between the so called adjoint sensitivity and the forward sensitivity method (FSM) introduced in Chap. 2. Adjoint sensitivity lies at the core of the well known 4-dimensional variational method known as 4D-VAR (Lewis et al. 2006, Chaps. 22–26).

We begin by expressing the adjoint sensitivity as the sum of products of forward sensitivity and the model forecast error where the sum is taken over the N time instances when observations are available. Understanding the impact of observations on adjoint sensitivity has long been a central issue in the study of 4D-VAR and numerous investigations have examined this question experimentally. Using the above-mentioned decomposition of adjoint sensitivity, we can provide a simple recipe for maximizing the impact of observations on adjoint sensitivity.

An investigation into this impact can best be understood by first considering a scalar constraint in discrete time.

3.1 On the Structure of Adjoint Sensitivity

Consider a discrete time, nonlinear scalar dynamical system described by

$$\mathbf{x}(k+1) = \mathbf{M}(\mathbf{x}(k), \alpha) \tag{3.1.1}$$

where, $\alpha \in \mathbb{R}^p$ is a vector of p parameters and $\mathbf{M} : \mathbb{R} \times \mathbb{R}^p \rightarrow \mathbb{R}$. Let $x(0)$ be the initial state for (3.1.1). Let $z(k_1), z(k_2), \ldots z(k_N)$ be the sequence of N scalar observations available at times

$$0 < k_1 < k_2, \ldots < k_N \tag{3.1.2}$$

where

$$z(k_1) = h(x(k_i)) + v(k_i) \tag{3.1.3}$$

© Springer International Publishing Switzerland 2017
S. Lakshmivarahan et al., *Forecast Error Correction using Dynamic Data Assimilation*, Springer Atmospheric Sciences, DOI 10.1007/978-3-319-39997-3_3

where $h : \mathbb{R} \to \mathbb{R}$ is the observation operator acts on the model variable and delivers the model counterpart to the observed physical quantity z. The observation error is modeled by Gaussian white noise with mean zero and variance σ^2, and is abbreviated as $N(0, \sigma^2)$.

The goal of the 4D-VAR method is to find the optimal value of the initial condition $x(0)$ and the parameter α such that the model trajectory matches the observations in the least square sense—minimization of the sum of squared errors. Mathematically, this estimation problem is often recast as a constrained minimization of an objective function $\mathbf{J} : \mathbb{R} \times \mathbb{R}^p \to \mathbb{R}$ given by

$$\mathbf{J}(\mathbf{c}) = \frac{1}{2} \sum_{i=1}^{N} \frac{(z(k_i) - h(x(k_i)))^2}{\sigma^2} \tag{3.1.4}$$

where the model state evolves according to (3.1.1) and the control vector $\mathbf{c} = (x(0), \alpha^T)^T \in \mathbb{R}^{1+p}$.

The fist step in minimizing (3.1.4) is to compute the gradient $\nabla_c J(\mathbf{c})$ of $J(\mathbf{c})$ with respect to $\mathbf{c}$ where

$$\nabla_{\mathbf{c}} \mathbf{J}(c) = \begin{bmatrix} \frac{\partial \mathbf{J}(\mathbf{c})}{\partial x(0)} \\ \cdots \\ \nabla_\alpha \mathbf{J}(\mathbf{c}) \end{bmatrix} \in \mathbb{R}^{1+p} \tag{3.1.5}$$

with $\frac{\partial \mathbf{J}(\mathbf{c})}{\partial x(0)} \in \mathbb{R}$ and $\nabla_\alpha \mathbf{J}(\mathbf{c}) = \left[\frac{\partial \mathbf{J}(\mathbf{c})}{\partial \alpha_1}, \frac{\partial \mathbf{J}(\mathbf{c})}{\partial \alpha_2}, \cdots \frac{\partial \mathbf{J}(\mathbf{c})}{\partial \alpha_p} \right]^T \in \mathbb{R}^p$.

3.1.1 Adjoint Method

The basic idea behind this method is equating two different forms of the gradient as a strategy for determination of an easily computable expression for the gradient in (3.1.5). Let $\delta \mathbf{c} \in \mathbb{R}^{1+p}$ be the perturbation of the control vector $\mathbf{c}$ and let $\delta \mathbf{J}$ be the corresponding first variation in $\mathbf{J}(\mathbf{c})$ induced by $\delta \mathbf{c}$. Then from first principles, we readily obtain

$$\delta \mathbf{J} = \langle \nabla_c \mathbf{J}, \delta \mathbf{c} \rangle \tag{3.1.6}$$

where $\langle \mathbf{a}, \mathbf{b} \rangle = \mathbf{a}^T \mathbf{b} = \sum_{i=0}^{n} a_i b_i$ is the standard inner product of two vectors $\mathbf{a}, \mathbf{b} \in \mathbb{R}^n$. In the following, we exploit two of the basic properties of the inner product, namely

Property 3.1 (Linearity).

$$\langle \mathbf{a} + \mathbf{b}, \mathbf{x} \rangle = \langle \mathbf{a}, x \rangle + \langle \mathbf{b}, \mathbf{x} \rangle$$

and

Property 3.2 (Adjoint Property).

$$\langle \mathbf{a}, \mathbf{A}\mathbf{b} \rangle = \langle \mathbf{A}^T\mathbf{a}, \mathbf{b} \rangle$$

where $\mathbf{a}, \mathbf{b}, \mathbf{x} \in \mathbb{R}^n$, $\mathbf{A} \in \mathbb{R}^{n \times n}$ and $\mathbf{A}^T$ is the transpose or adjoint of $\mathbf{A} \in \mathbb{R}^{n \times n}$. See Friedman (1956, Chap. 1) for elaboration on these properties.

Taking the first variation of (3.1.4) it follows that

$$\delta \mathbf{J} = -\sum_{i=1}^{N} \langle e_F(k_i), \delta h\,(x(k_i)) \rangle \tag{3.1.7}$$

where

$$e_F(k_i) = \frac{1}{\sigma^2}\,[z(k_i) - h\,(x(k_i))] \tag{3.1.8}$$

is the normalized nonlinear model forecast error at time k_i,

$$\delta \mathbf{h}\,(x) = \mathbf{D}_x(h)\delta \mathbf{x}. \tag{3.1.9}$$

and σ^2 is the error variance.

Since $e_F(k_i)$ and $\delta h\,(x(k_i))$ are scalars, the use of inner product notation to express the simple product in (3.1.7) looks superfluous at first sight. However, since δx is the sum of contributions from variations of $\delta x(0)$ (which is scalar) and $\delta \alpha$ (which is a vector of size p), this notation is appropriate as will soon become evident.

Substituting (3.1.9) in (3.1.7) and using the adjoint Property 3.2, we get

$$\delta \mathbf{J} = -\sum_{i=1}^{N} \langle \mathbf{D}_{x(k_i)}^T(h)e_F(k_i), \delta x(k_i) \rangle \tag{3.1.10}$$

where $\mathbf{D}_x(h)$ is a scalar in this case and the transpose of a scalar is equal to itself.

But from the model equation (3.1.1) and from Sect. 2.2, it follows that

$$\delta x(k_i) = \mathbf{D}_{x(0)}(x(i))\delta x(0) + \mathbf{D}_\alpha(M(x(i), \alpha))\delta \alpha \tag{3.1.11}$$

where recall that

$$\mathbf{D}_{x(0)}(x(i)) = u_1(i) = \frac{\partial x(i)}{\partial x(0)} \in \mathbb{R}, \quad \text{and} \tag{3.1.12}$$

$$\mathbf{D}_\alpha(x(i)) = v_1(i) = \frac{\partial x(i)}{\partial \alpha} \in \mathbb{R}^p \tag{3.1.13}$$

are the forward sensitivities of $x(i)$ with respect to the initial condition $x(0) \in \mathbb{R}$ and the parameter $\alpha \in \mathbb{R}^p$, respectively.

Substituting (3.1.11) into (3.1.10) and using the Property 3.1, it follows that

$$\delta \mathbf{J} = - \sum_{i=1}^{N} \langle \mathbf{D}_{x(k_i)}^{T}(h) e_F(k_i), \mathbf{D}_{x(0)}(x(k_i)) \, \delta x(0) \rangle$$

$$- \sum_{i=1}^{N} \langle \mathbf{D}_{x(k_i)}^{T}(h) e_F(k_i), \mathbf{D}_{\alpha}(x(k_i)) \delta \alpha \rangle. \tag{3.1.14}$$

Using the adjoint property again, we get

$$\delta J = \text{Term}_I + \text{Term}_{II} \tag{3.1.15}$$

where

$$\text{Term}_I = - \sum_{i=1}^{N} \langle \mathbf{D}_{x(0)}^{T}(x(k_i)) \eta_F(k_i), \delta x(0) \rangle \tag{3.1.16}$$

and

$$\text{Term}_{II} = - \sum_{i=1}^{N} \langle \mathbf{D}_{\alpha}^{T}(x(k_i)) \eta_F(k_i), \delta \alpha \rangle \tag{3.1.17}$$

where

$$\eta_F(k_i) = \mathbf{D}_{x(k_i)}^{T}(h) e_F(k_i) \tag{3.1.18}$$

is the transformed and normalized forecast error as observed from the model space. From first principles, we get

$$\text{Term}_I = \langle \nabla_{x(0)} \mathbf{J}, \delta x(0) \rangle. \tag{3.1.19}$$

Since $\langle \mathbf{a}, \mathbf{x} \rangle = \langle \mathbf{b}, \mathbf{x} \rangle$ for all $\mathbf{x}$ implies that $\mathbf{a} = \mathbf{b}$, and applying this equality to (3.1.16) and (3.1.19) the fine structure of the adjoint sensitivity can be expressed as

$$\frac{\partial \mathbf{J}}{\partial x(0)} = \nabla_{x(0)} \mathbf{J} = - \sum_{i=1}^{N} \mathbf{D}_{x(0)}^{T}(x(k_i)) \, \eta_F(k_i), \tag{3.1.20}$$

which is the sum of products of the transpose of forward sensitivity, $\mathbf{D}_{x(0)} x(k_i)$, with the transformed forecast error, $\eta_F(k_i)$. Similarly

$$\text{Term}_{II} = \langle \nabla_{\alpha} \mathbf{J}, \delta \alpha \rangle. \tag{3.1.21}$$

By equating the right hand sides of (3.1.17) and (3.1.21), it follows that

$$\frac{\partial J}{\partial \alpha} = \nabla_\alpha J = -\sum_{i=1}^{N} D_\alpha^T \left(x(k_i) \right) \eta_F(k_i) \tag{3.1.22}$$

where $D_\alpha^T (x)$ and $\nabla_\alpha J$ are vectors in $\mathbb{R}^p$. And as before, clearly (3.1.22) expresses $\nabla_\alpha J$ as the sum of the products of forward sensitivity $D_\alpha^T (x(k_i))$ with respect to α with the transformed forecast error $\eta_F(k_i)$.

3.1.2   Computing $\nabla_{x(0)} J$ in (3.1.20)

We now proceed to describe an efficient algorithm for computing the gradient of J in (3.1.20). Recall from Problem 2.5 in Chap. 2 that $A(k) = D_{x(k)}(M)$, the model Jacobian at time k and

$$D_{x(0)} \left(x(k) \right) \prod_{i=0}^{k-1}{}' A(i) = A(k-1)A(k-2)\ldots A(1)A(0)$$

$$= A(k-1:0) \tag{3.1.23}$$

and

$$D_{x(0)}^T \left(x(k) \right) \prod_{i=0}^{k-1} A^T(i) = A^T(0)A^T(1)\ldots A^T(k-2)A^T(k-1)$$

$$= A^T(0:k-1). \tag{3.1.24}$$

Clearly, $D_{x(0)} (x(k))$ has two interpretations: (1) it is the reverse product of Jacobians of the model map M in (3.1.1) along the forecast trajectory, and (2) it is the forward sensitivity of the solution $x(k)$ with respect to the initial conditions. In the adjoint method which is the topic of this section, we use interpretation (1); the FSM-based interpretation is used in the section below.

Let $OT = \{k_1, k_2, \ldots, k_N\}$ be an ordered set of N increasing integers that represents the times when observations are available for assimilation. Define $\overline{N} = k_N$ be the integer corresponding to the last observation time. Let $\overline{I} = \{1, 2, \ldots, \overline{N}\}$ denote the set of all non-negative integers in the interval $[0, \overline{N}]$. Clearly, $OT \subset \overline{I}$, that is, OT is a subset of $\overline{I}$.

Recall that the transformed forecast errors $\{\eta_F(k_i) | 1 \leq i \leq N\}$ are defined on the set OT. We now extend the definition of forecast errors to the larger set $\overline{I}$. To this end, let

$$\overline{\eta}_F(k) = \delta_{k,k_i} \eta_F(k_i) \tag{3.1.25}$$

where

$$
\delta_{k,k_i} = \begin{cases} 1 & \text{if } k = k_i \\ 0 & \text{otherwise,} \end{cases}
$$

where $k \in \bar{I}$ and $k_i \in OT$. Stated in words, $\bar{\eta}_F(k)$ is defined on $\bar{I}$ and it coincides with $\eta_F(k_i)$ on OT and zero outside of OT in $\bar{I}$.

Using this form of $\bar{\eta}_F$, we can rewrite (3.1.20) as

$$
\nabla_{x(0)}\mathbf{J} = -\sum_{i=1}^{\bar{N}} \mathbf{D}_{x(0)}^T (x(i))\, \bar{\eta}_F(i) \tag{3.1.26}
$$

where the sum on the right-hand side is over all integers $1 \le i \le \bar{N}$ of the integers in the subset of OT. The rationale for (3.1.26) is that the expression can be easily computed using a backward recursion starting from the last index $\bar{N}$ as

$$
\begin{cases} \lambda(N) = \bar{\eta}_F(\bar{N}) = \eta_F(k_N), \text{ and} \\ \lambda(i) = A^T(k)\lambda(i+1) + \bar{\eta}(i). \end{cases} \tag{3.1.27}
$$

It can be verified (see Problem 3.1) that

$$
\lambda(0) = -\nabla_{x(0)}\mathbf{J} = \sum_{i=0}^{\bar{N}} \left[\prod_{q=0}^{i} \mathbf{A}^T(q) \right] \bar{\eta}_F(i). \tag{3.1.28}
$$

Notice that while these are $(1 + \bar{N})$ terms in the summation found on the right-hand side of (3.1.28), since only N of the $\bar{\eta}_F(i)$ are non-zero; thus in effect there are only N-terms in (3.1.28) as should be expected from the form of (3.1.20). The text of the complete algorithm for computing $\nabla_{x(0)}\mathbf{J}$ is given in Algorithm 3.1

Algorithm 3.1 Algorithm for computing the gradient $\nabla_{x(0)}J$ using the first-order adjoint method.

Step 1: Specify the values of initial condition $x(0)$ and parameters.
Step 2: Compute the forecast trajectory $x(k), k \ge 0$ using the model equation (3.1.1)
Step 3: Given the observations $z(k_i), 1 \le i \le N$ in (3.1.3) compute the forecast error $e_F(k_i)$ using (3.1.8)
Step 4: Compute the Jacobian $\mathbf{D}_x(h)$ of the observation operator h and evaluate $\mathbf{D}_{x(k_i)}(h)$ at observation times k_i using $x(k)$ in Step 1.
Step 5: Compute the transformed forecast errors $\eta_F(k_i), 1 \le i \le N$
Step 6: Set up and solve the backward recurrence in (3.1.27) to obtain the gradient $\nabla_{x(0)}J$ as given by (3.1.28)

As an example, if $k_1 = 1, k_2 = 3$ and $k_3 = 5$, then $N = 3, \overline{N} = 5$ and

$$- \nabla_{x(0)} J = \left[\mathbf{A}^T (0) \right] \eta_F (1) + \left[\mathbf{A}^T (0 : 2) \right] \eta_F (3) + \left[\mathbf{A}^T (0 : 4) \right] \eta_F (5), \qquad (3.1.29)$$

where $\mathbf{A}^T (0 : j)$ is defined in (3.1.24).

3.1.3 Computation of $\nabla_\alpha J$ in (3.1.22)

Following the development above, we start by rewriting (3.1.22) as

$$\nabla_\alpha (\mathbf{J}) = - \sum_{i=1}^{N} \mathbf{D}_\alpha^T (x(i)) \, \overline{\eta}_F (i). \qquad (3.1.30)$$

Since $\prod'$ denotes product in the decreasing order of indices based on results in Problems 2.3 and 2.4 in Chap. 2, we get

$$\mathbf{D}_\alpha (x(i)) = \sum_{k=0}^{i-1} \left[\prod_{q=k+1}^{i-1} \mathbf{A}(q) \right] \mathbf{B}(k) = \sum_{k=0}^{i-1} \mathbf{A}(i - 1 : k + 1) \mathbf{B}(k). \qquad (3.1.31)$$

Hence,

$$\mathbf{D}_\alpha^T (x(i)) = \sum_{k=0}^{i-1} \mathbf{B}^T (k) A^T (k + 1 : i - 1). \qquad (3.1.32)$$

Consequently,

$$- \nabla_\alpha (\mathbf{J}) = \sum_{i=1}^{N} \left(\sum_{k=0}^{i-1} \mathbf{B}^T (k) A^T (k + 1 : i - 1) \right) \overline{\eta}_F (i). \qquad (3.1.33)$$

This expression for $\overline{N} = 4$ is given by

$$- \nabla_\alpha (\mathbf{J}) = \mathbf{B}^T (0) \overline{\eta}_F (1)$$
$$+ \left[\mathbf{B}^T (0) A^T (1) + \mathbf{B}^T (1) \right] \overline{\eta}_F (2)$$
$$+ \left[\mathbf{B}^T (0) A^T (1) A^T (2) + \mathbf{B}^T (1) A^T (2) + \mathbf{B}^T (2) \right] \overline{\eta}_F (3)$$
$$+ \left[\mathbf{B}^T (0) A^T (1) A^T (2) A^T (3) + \mathbf{B}^T (1) A^T (2) A^T (3) \right.$$
$$\left. + \mathbf{B}^T (2) A^T (3) + \mathbf{B}^T (3) \right] \overline{\eta}_F (4),$$

which can be rewritten as

$$
\begin{aligned}
-\nabla_\alpha(\mathbf{J}) = \mathbf{B}^T(0) &\left[\overline{\eta}_F(1) + \mathbf{A}^T(1)\overline{\eta}_F(2) + \mathbf{A}^T(1)\mathbf{A}^T(2)\overline{\eta}_F(3)\right. \\
&\left.+\mathbf{A}^T(1)\mathbf{A}^T(2)\mathbf{A}^T(3)\overline{\eta}_F(4)\right] \\
&+\mathbf{B}^T(1)\left[\overline{\eta}_F(2) + \mathbf{A}^T(2)\overline{\eta}_F(3) + \mathbf{A}^T(2)\mathbf{A}^T(3)\overline{\eta}_F(4)\right] \\
&+\mathbf{B}^T(2)\left[\overline{\eta}_F(3) + \mathbf{A}^T(3)\overline{\eta}_F(4)\right] \\
&+\mathbf{B}^T(3)\left[\overline{\eta}_F(4)\right].
\end{aligned}
\tag{3.1.34}
$$

Generalizing this we obtain

$$
-\nabla_\alpha(J) = \sum_{k=0}^{N-1} \mathbf{B}^T(k) \left(\sum_{j=k}^{N-1} \mathbf{A}^T(k+1:j)\overline{\eta}_F(j+1) \right).
\tag{3.1.35}
$$

The text of the complete algorithm for computing $\nabla_\alpha J$ is given in Algorithm 3.2

3.1.4   4D-VAR method

Once gradient $\nabla_c\,(\mathbf{J}(\mathbf{c}))$ is made available, we move to iteratively minimizing $\mathbf{J}(\mathbf{c})$ using one of many well known methods including the gradient method, conjugate gradient method or quasi-Newton methods. These methods are described in detail in Lewis et al. (2006, Chaps. 10–12). For completeness we indicate the framework of a minimization strategy in Algorithm 3.3.

Algorithm 3.2 Algorithm for computing the adjoint sensitivity with respect to the parameter α

> **for** $k = 1$ to $\overline{N}$ **do**
> $\quad \lambda(N) = \overline{\eta}_F(N)$
> $\quad$ **for** $j = \overline{N} - 1 : k$ **do**
> $\quad\quad \lambda(j) = \mathbf{A}^T(j)\lambda(j+1) + \overline{\eta}_F(j)$
> $\quad$ **end for**
> $\quad \overline{\lambda}(k) = \mathbf{B}^T(k-1)\lambda(k)$
> **end for**
> *sum* $= 0$
> **for** $k = 1$ to $\overline{N}$ **do**
> $\quad$ *sum* $=$ *sum* $+ \overline{\lambda}(k)$
> **end for**
> $\nabla_c(\mathbf{J}) = -sum$

Algorithm 3.3 A framework minimization to be used in 4D-VAR method.

> *Step 1:* Pick **c** the initial value of control.
> *Step 2:* Compute $\nabla_c \mathbf{J}(\mathbf{c})$ as prescribed in this section
> *Step 3:* Define $\mathbf{c}_{new} = \mathbf{c}_{old} - \beta \nabla_c \mathbf{J}(\mathbf{c})$, where β is the step length whose choice depends on the algorithm used
> *Step 4:* Check the exit condition for convergence. If satisfied, exit, else, continue with Step 2

3.2   On the Relation Between FSM and 4D-VAR

Let $\boldsymbol{\zeta} = \delta \mathbf{c} = \left(\delta x(0), \delta \alpha^T \right)^T \in \mathbb{R}^{1+p}$ be the increment in the current control **c**. Then, from first principles, we define

$$\delta \mathbf{J} = \mathbf{J}(\mathbf{c} + \boldsymbol{\zeta}) - \mathbf{J}(\mathbf{c}). \tag{3.2.1}$$

From (3.1.14), we then obtain (using adjoint property)

$$\delta \mathbf{J} = - \sum_{i=1}^{N} \langle e_F(k_i), H(k_i)\boldsymbol{\zeta} \rangle, \tag{3.2.2}$$

where

$$\begin{cases} \boldsymbol{\zeta} = \left(\delta x^T(0), \delta \alpha^T \right)^T, \\ \mathbf{H}(k_i) = [H_1(k_i), H_2(k_i)] \in \mathbb{R}^{1+p}, \\ H_1(k_i) = \mathbf{D}_{x(k_i)}(h)\mathbf{D}_{x(0)}(x(k_i)), \\ H_2(k_i) = \mathbf{D}_{x(k_i)}(h)\mathbf{D}_\alpha(x(k_i)). \end{cases} \tag{3.2.3}$$

Also, recall that since

$$\delta \mathbf{J} = \langle \nabla_c \mathbf{J}, \boldsymbol{\zeta} \rangle \tag{3.2.4}$$

and $\boldsymbol{\zeta}$ is arbitrary, at the minimum $\nabla_c \mathbf{J} = 0$ implies $\delta \mathbf{J}$ is also zero. From this and from (3.2.1)–(3.2.2), it immediately follows that $\delta \mathbf{J}$ must vanish at the minimum, which in turn implies that

$$\mathbf{H}(k_i)\boldsymbol{\zeta} = e_F(k_i), 1 \le i \le N \tag{3.2.5}$$

Defining

$$\mathbf{H} = \begin{bmatrix} H(k_1) \\ H(k_2) \\ \vdots \\ H(k_n) \end{bmatrix} \in \mathbb{R}^{N \times (1+p)}, \quad \mathbf{e}_F = \begin{bmatrix} e_F(k_1) \\ e_F(k_2) \\ \vdots \\ e_F(k_n) \end{bmatrix} \in \mathbb{R}^N$$

(3.2.5) takes the form

$$\mathbf{H}\zeta = \mathbf{e}_F \qquad (3.2.6)$$

which is the defining equation for the FSM described in Chap. 2.

3.3 Investigation of the Impact of Observations Using 4D-VAR and FSM

As claimed in the introductory chapter of this book, the evolution of sensitivities of the model output to elements of control can be used to place observations (in space and/or time) where they best serve the goal of correcting control—corrections that minimize the sum of squared differences between the forecast and observations. As before, we use the air-sea interaction dynamics to demonstrate the adequacy, or inadequacy of observations to correct control. We use the discrete form of the constraint given by

$$x_{k+1} = \left(1 - \frac{\nu}{10}\right)^k (x_0 - \theta) + \theta$$

where x is the air temperature at time $t = k\Delta t$, $k = 0, 1, 2, \ldots$, $\Delta t = 1$ h, ν is the turbulent transfer coefficient, θ the sea surface temperature and x_0 the initial air temperature at $t = 0$ ($k = 0$). We replace the transfer coefficient c by $\nu/10$ as a form of preconditioning that leads to faster convergence of the optimization process. The sensitivities of prediction to elements of control for this problem have been discussed and exhibited through graphics in Sect. 2.1. The form of constraint and sensitivities were based on continuous form in this earlier presentation. A cursory examination of the sensitivity graphics (Figs. 2.2, 2.3, and 2.4) makes it clear that a set of observations limited to the later times, say beyond 10 h, will make it difficult to correct errors in the initial condition and parameter ν. Errors in the prediction at these later times stem mostly from uncertainty in θ.

On the other hand, a forecast error in prediction at an early time, say $t = 1$ or 2 h, can be used to correct the initial condition as well as the parameter ν since errors in these controls will significantly contribute to the prediction error. This line of argument leads to a strategy for placement of observations to correct all elements of control. Placement of some observations will favor corrections to a given element or elements. Other controls may not be correctable from observation at these locations. Thus, judicious placement of observations such that key elements of control are correctable is the important strategy.

We conduct two numerical experiments using both 4D-VAR and FSM. In Experiment I, four observations are limited to later times: $t = 15, 16, 17$ and 18 h. In Experiment II, a separate set of four observations are spread over the forecast horizon such that all elements of control have at least one placement of observations where the forecast is sensitive to that control. Here we place observations at $t = 1, 2, 17$ and 18 h.

In these experiments, the observations are created by using the discrete form of the constraint with controls $(x_0, \theta, v) = (1\,°C, 11\,°C, 2.5)$. The guess control for the data assimilation experiments is given by $(x_0', \theta', v') = (2\,°C, 10\,°C, 3.5)$. Let the times be represented by t_1, t_2, t_3, and t_4, where the associated indices in the constraint are $k = 15, 16, 17$, and 18, respectively, for Experiment I; $k = 1, 2, 17$, and 18 for Experiment II.

In both experiments, the observations are the observed air temperatures denoted by z_1, z_2, z_3 and z_4, where the k indices are the same as those associated with observation times.

3.3.1   Experiments

The functional to be minimized in the 4D-VAR experiments is given by

$$J(x_0, \theta, v) = \sum_{k=1}^{4} [z_i - x_i]^2 .$$

The 4D-VAR method requires we first find the gradient of J, ∇J, using the guess control and its associated forecast to the times t_1, t_2, t_3, and t_4. As shown in the main body of the text, the components of the ∇J are products of the sensitivity and forecast error. In this case where $h(x) = x$, these components are

$$\frac{\partial J}{\partial x_0} = \sum_{k=1}^{4} [z_k - x_k] \frac{\partial x_k}{\partial x_0},$$

$$\frac{\partial J}{\partial \theta} = \sum_{k=1}^{4} [z_k - x_k] \frac{\partial x_k}{\partial \theta},$$

$$\frac{\partial J}{\partial v} = \sum_{k=1}^{4} [z_k - x_k] \frac{\partial x_k}{\partial v}.$$

Once the components of ∇J are calculated, the unit vector in the $-\nabla J$ direction is found. A step size is then found along this direction to iterate toward the minimum of $J(x_0, \theta, v)$. The step size is the estimate of distance along $-\nabla J$ where J has the smallest value. We use the parabolic profile assumption to find the step size. The slope of J at the start point (the operating point on control), the value of J at the starting point, and another value of J at an arbitrarily chosen distance along the $-\nabla J$ direction are the conditions that determine the parabolic profile and the point at the minimum value of the parabola determines the step size.

Table 3.1 Results for experiment I (4D-VAR)

Control			$-\nabla J$			Adjustments		
x_0	θ	v	$-\frac{\partial J}{\partial x_0}$	$-\frac{\partial J}{\partial \theta}$	$-\frac{\partial J}{\partial v}$	Δx_0	$\Delta \theta$	Δv
Step 1: J decreases from 1.741 to 0.000758								
2.0	10.0	3.0	0.001	3.720	0.206	0.0028	0.9326	0.0513
Step 2: J decreases from 0.000758 to 0.000757								
2.003	10.933	3.051	−8.7e-5	−1.8e-3	−1.5e-3	−3e-5	−6e-4	−6e-4

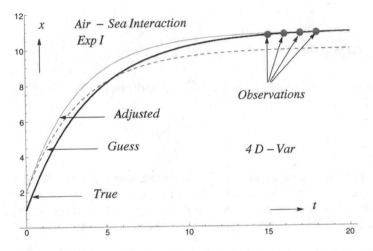

Fig. 3.1 4D-VAR adjustment when observations are in the 'saturated zone' (where temperature changes with time are minuscule)

Examination of results in Table 3.1 makes it clear that the value of the cost function for Experiment I is decreased significantly in two iterations—from 1.741 to 0.0008. However, only θ receives meaningful adjustment. Although the initial condition x_0 and parameter v only receive minute adjustment, the fit of model to observations is excellent as shown in Fig. 3.1. Even though the process reduces the value of the functional significantly and fits the observations, this result is deceptively poor. As noted in Fig. 3.1, the adjusted forecast at earlier times in the range of $0 - 10\,h$ exhibits serious error due to incorrect x_0 and v. Even prior to making an adjustments to control in step 1 of Experiment I, note that the sign of the negative gradient of J with respect to x_0 and v leads to adjustments in the wrong directions for both of these elements.

As noted in both steps 1 and 2 of Table 3.1, the gradient of J with respect to the x_0 and v are extremely small—especially in step 2. These extremely small magnitudes essentially indicate that the structure of J in the vicinity of the operating point is flat along these two directions. In essence, the search for the minimum has become "marooned" in the "flatland" of these elements. This feature of 4D-VAR can sometimes be ameliorated through preconditioning, a form of scaling and alternative

Table 3.2 Results for experiment II (4D-VAR)

Control			$-\nabla J$			Adjustments		
x_0	θ	ν	$-\frac{\partial J}{\partial x_0}$	$-\frac{\partial J}{\partial \theta}$	$-\frac{\partial J}{\partial \nu}$	Δx_0	$\Delta\theta$	$\Delta\nu$
Step 1: J decreases from 1.741 to 0.000758								
2.0	10.0	3.0	−0.972	1.267	−1.435	−0.612	0.798	−0.903
Step 2: J decreases from 0.000758 to 0.000757								
1.388	10.797	2.098	0.403	0.771	1.064	0.087	0.166	0.217
Step 3								
1.475	10.964	2.314						

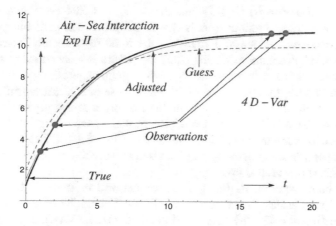

Fig. 3.2 4D-VAR adjustment when observations are spread over the time domain

representation of the variables. In this case, however, the preconditioning introduced ν as a function of the exchange coefficient c was inconsequential.

Results for Experiment II of the 4D-VAR data assimilation are displayed in Table 3.2. In the first step, it is clear that the signs of the negative gradient of J with respect to all elements of control lead to adjustment in the optimal direction (negative for x_0 and ν and positive θ). Indeed, the adjustment through two steps leads to controls where the associated forecast fits the observations well—at both the early and later times (Fig. 3.2). In short, the structure of J in the vicinity of the operating point gives rise to negative gradients that lead to meaningful adjustments via steepest descent strategy.

It is instructive to examine the search for the minimum of the cost functional in this relatively simple example. Let us look at the first step of Experiment II. The unit vector along the negative gradient direction (call it the r direction) is (−0.423, 0.590, −0.668) for components x_0, θ, and V, respectively. Division of the $-\nabla J$ by its absolute value produces this unit vector. The control vector at $r = 0$ is the guess control for the first step. As r increases, J decreases. At what value of r does the value of J reach its minimum? In this example, the values of J along r can be calculated by "brute force". That is, be sequentially increasing r and finding the control at that point (the value of r multiplied by the unit vector plus the control at $r = 0$), make a forecast using the computed control, and substitute the associated forecast into the cost function. As you will note, a separate forecast is required at each value of r. This requirement is computationally expensive for more complex models. And indeed, no matter the strategy used to estimate the step size, at least one forecast is required. Brute force determination of the step size is not advised. As mentioned earlier, we fit a parabola to the cost function along the r direction. The conditions used to find the coefficients of $J = Ar^2 + Br + C$ are as follows: value of J at $r = 0$, its value at an arbitrary chosen value or r, and its slope, $\frac{dJ}{dr}$ at $r = 0$ (equal to the product of $-\nabla J$ at $r = 0$ and the unit vector). For Step 1 of experiment II, $\frac{dJ}{dr}\big|_{r=0} = -2.15, J(r = 0) = 1.56$, and $J(r = 1) = 0.204$—A, B and C are found from these conditions and the plot is shown in Fig. 3.1. The parabolic profile is plotted over the exact values of J found by the brute force method discussed above. As noted, the J minimum for the parabolic profile is reached at $r = 1.35$ whereas the actual minimum is found at $r = 1.25$—a good match.

As shown in Chap. 2, the foundation of the FSM is use of forecast sensitivity to adjust control. In this demonstration, we employ first-order sensitivity analysis in the iterative mode. That is, the guess control is adjusted by requiring that the sum of squared residuals is minimized (residual defined in Chap. 2, Sect. 2.4). Using the adjusted control as a second operating point, revised sensitivities are calculated and again used to adjust this second operating point.

Results for both Experiments I and II are summarized in Tables 3.3 and 3.4. There we note similarities to the results of 4D-VAR assimilation as found in Tables 3.1 (Experiment I: 4D-VAR) and 3.2 (Experiment II: 4D-VAR). In particular, the sea surface temperature control θ is well adjusted by both 4D-VAR and FSM in Experiment I, but the x_0 and ν are "poorly adjusted"; that is, after adjustment

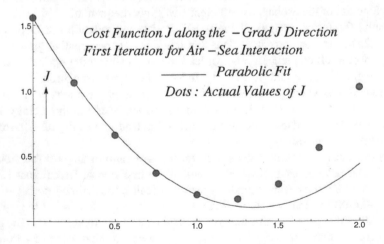

Fig. 3.3 *Dots* are values of the functional along the steepest-descent direction and the parabolic curve determines the step size that estimates the minimum value of the functional

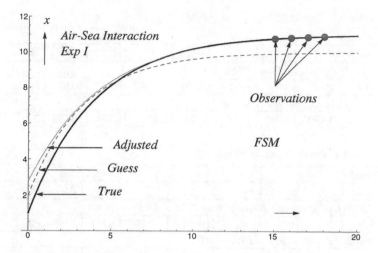

Fig. 3.4 FSM adjustment when observations are in the 'saturated zone' (where temperature changes with time are minuscule)

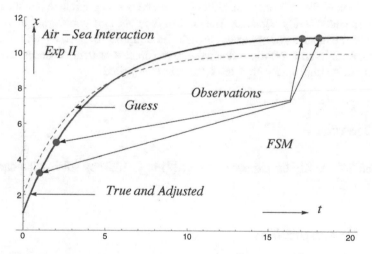

Fig. 3.5 FSM adjustment when observations are spread over the time domain

they remain far from the true control. Yet, the fit of the forecast to the observations is excellent in both forms of data assimilation. And, indeed, the two assimilation procedures produce forecasts that exhibit error at the early times in response to the poor adjustment for x_0 and v. For Experiment II, FSM produces an excellent adjustment to control and 4D-VAR, after two iterations, is very good.

The experiments convincingly illustrate the value of knowledge of sensitivity evaluation in positioning observations. The results also indicate that caution should be used in relying on the reduction in value of the cost function alone. In short,

Table 3.3 Results for experiment I (FSM)

Control			$-\nabla J$			Adjustments		
x_0	θ	v	$-\frac{\partial J}{\partial x_0}$	$-\frac{\partial J}{\partial \theta}$	$-\frac{\partial J}{\partial v}$	Δx_0	$\Delta \theta$	Δv
Step 1: J decreases from 1.741 to 0.053								
2.0	10.0	3.0	−1.408	0.992	−0.948	0.562	10.992	2.052
Step 2: J decreases from 0.053 to 0.003								
0.592	10.992	2.052	2.179	0.002	0.214	2.771	10.994	2.266

Table 3.4 Results for experiment II (FSM)

Control			$-\nabla J$			Adjustments		
x_0	θ	v	$-\frac{\partial J}{\partial x_0}$	$-\frac{\partial J}{\partial \theta}$	$-\frac{\partial J}{\partial v}$	Δx_0	$\Delta \theta$	Δv
Step 1: Cost function J decreases from 1.556 to 0.169								
2.0	10.0	3.0	−0.839	0.989	−0.887	1.161	10.989	2.113
Step 2: J decreases from 0.169 to 0.002								
1.161	10.989	2.113	−0.139	−0.030	0.385	1.022	10.959	2.498

minimization of the cost function, although achieved, is no guarantee that the control vector has been ideally adjusted. The control element that is most sensitive to the forecast at the location of observations will generally be adjusted meaningfully, but other controls that have little influence on the forecast at these observation points are likely to be improperly adjusted or remain unadjusted.

3.4   Exercises

Problem 3.1. Verify the expressions for $\lambda(0)$ in (3.1.28) by solving the backward recurrence (3.1.27)

3.4.1   Demonstrations

Demonstration: Inertial Motion in the Ocean (Ocean Inertia I)

The classic study of oscillatory motion in the Baltic Sea is used as our guide in the experiment—a study conducted in August 1933 by Gustafson and Kullenberg (1936). As stated by Sverdrup et al. (1942) in their review of this

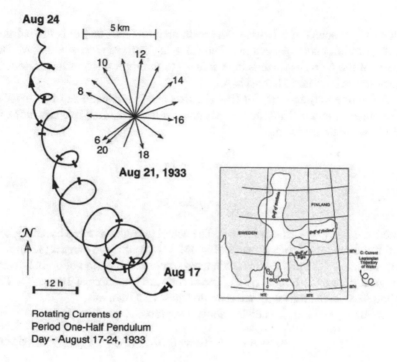

Fig. IM.1 Trajectory of Baltic sea water (*left*) in the presence of rotating current (*upper center*)

study, an impulsive action of a squall line was likely responsible for initiating this observed oscillatory motion superimposed on a base flow. The trajectory of water movement for this case is shown in Fig. IM.1 Fortuitously, the Coriolis force was dominant in this case and the typically strong horizontal pressure gradient force was weak. Consequently, the governing equations for the horizontal motion of water is given by

$$\frac{\partial u}{\partial t} - fv = 0$$

$$\frac{\partial v}{\partial t} + fu = 0 \tag{IM.1}$$

$$\frac{\partial u}{\partial x} + \frac{\partial v}{\partial y} = 0$$

where u, v are the standard velocity components of the water along the x (toward east) and y (toward north) directions respectively, and t denotes the time. The Coriolis parameter is

$$f = 2\Omega_E \sin \Phi \tag{IM.2}$$

where Ω_E is speed of rotation of the earth about its axis and Φ is the latitude of the location under observation. The value of f is fixed with $\Phi = 53°N$. The action of the Coriolis force is *cum sole* — to the right (left) of the current in the northern (southern) hemisphere.

A solution $u(t)$ and $v(t)$ for (IM.1) exists and is independent of the spatial coordinates x and y. Thus, adding a friction term to the first two equations in (IM.1) we get our base model as

$$\frac{du}{dt} = fv - ku$$
$$\frac{dv}{dt} = -fu - kv$$

(IM.3)

where the friction coefficient $k > 0$. The actual current observed by Gustafson and Kullenberg (1936) is shown in Fig. IM.1. Note that the current (vectors in upper center of the plot) exhibit a damped motion with period of $\sim 14\,h$ (from $6\,h$ local time to $20\,h$ local time). The speeds are the order of $10\,cm\,s^{-1}$. The "short bar" on the trajectory are recorded at $\sim 12\,h$ intervals.

Define $\mathbf{V}_I = (u, v)^T$. Then the above base model equations become

$$\dot{\mathbf{V}}_I = \mathbf{A}\mathbf{V}_I, \quad \mathbf{V}_I(0) = (u(0), v(0))^T$$

(IM.4)

where

$$\mathbf{A} = \begin{bmatrix} -k & f \\ -f & -k \end{bmatrix}$$

(IM.5)

It can be verified that the eigenvalues $\mathbf{A}$ are obtained as the solution of the polynomial equation

$$0 = \begin{bmatrix} -k - \lambda & f \\ -f & -k - \lambda \end{bmatrix} = (k + \lambda)^2 + f^2,$$

and they are given by

$$\lambda_{1,2} = -k \pm if.$$

(IM.6)

The solution of (IM.4) represents the instantaneous velocity and is given by

$$\mathbf{V}(t) = e^{\mathbf{A}t}\,\mathbf{V}(0)$$

(IM.7)

where $e^{\mathbf{A}}$ is given by the power series

$$e^{\mathbf{A}} = \mathbf{I} + \mathbf{A} + \frac{\mathbf{A}^2}{2!} + \frac{\mathbf{A}^3}{3!} + \cdots.$$

(IM.8)

We can readily split $\mathbf{A}$ as a sum of $\mathbf{A}_1$ and $\mathbf{A}_2$ where

$$\mathbf{A}_1 = \begin{bmatrix} -k & 0 \\ 0 & -k \end{bmatrix}, \mathbf{A}_2 = \begin{bmatrix} 0 & f \\ -f & 0 \end{bmatrix} \text{ and } \mathbf{A} = \mathbf{A}_1 + \mathbf{A}_2.$$

It can also be verified that

$$\mathbf{A}_1 \mathbf{A}_2 = \mathbf{A}_2 \mathbf{A}_1. \tag{IM.9}$$

Since $e^{\mathbf{A}+\mathbf{B}} = e^{\mathbf{A}} e^{\mathbf{B}}$ where $\mathbf{AB} = \mathbf{BA}$, it is immediate that in view of (IM.9)

$$e^{\mathbf{A}t} = e^{(\mathbf{A}_1 + \mathbf{A}_2)t} = e^{\mathbf{A}_1 t} e^{\mathbf{A}_2 t}. \tag{IM.10}$$

Using the definition in (IM.8), and the series expansion for e^x, $\sin x$, $\cos x$, it can be verified that

$$e^{\mathbf{A}_1 t} = \begin{bmatrix} e^{-kt} & 0 \\ 0 & e^{-kt} \end{bmatrix} \tag{IM.11}$$

and

$$e^{\mathbf{A}_2 t} = \begin{bmatrix} \cos ft & \sin ft \\ -\sin ft & \cos ft \end{bmatrix} \tag{IM.12}$$

Substituting (IM.10), (IM.11), and (IM.12) in (IM.7), we get explicit expression for the velocity $\mathbf{V}_I(t)$ as

$$\mathbf{V}_I(t) = e^{-kt} \begin{bmatrix} u(0) \cos ft + v(0) \sin ft \\ -u(0) \sin ft + v(0) \cos ft \end{bmatrix} \tag{IM.13}$$

which is oscillatory motion with exponential damping.

Let the prevailing constant background velocity of water motion be given by

$$\mathbf{V}_B = (\mathbf{V}_{1,B}, \mathbf{V}_{2,B})^T = (U, V)^{[1]} \tag{IM.14}$$

Hence a typical water parcel moves with a velocity $\mathbf{V}$ given by

$$\mathbf{V}(t) = \mathbf{V}_I(t) + \mathbf{V}_B \tag{IM.15}$$

In compound form, (IM.15) becomes

$$\begin{aligned} u(t) &= V_{1,B} + e^{-kt} [u(0) \cos ft + v(0) \sin ft] \\ v(t) &= V_{2,B} + e^{-kt} [-u(0) \sin ft + v(0) \cos ft] \end{aligned} \tag{IM.16}$$

---

[1] In demonstration: Ocean inertia II that follows, the background velocity is referred to as U, V.

Let $\mathbf{x}(t) = (x(t), y(t))^T$ be the instantaneous position of a typical water parcel. Then

$$\mathbf{x}(t) = \int_0^t \mathbf{V}_B ds + \int_0^t \mathbf{V}_I(s) ds \tag{IM.17}$$

$$= \mathbf{V}_B t + \left[\int_0^t e^{As} ds \right] \mathbf{V}(0) \tag{IM.18}$$

$$= \mathbf{V}_B t + \mathbf{A}^{-1} \left[e^{As} \right]_0^t \mathbf{V}(0)$$

$$= \mathbf{V}_B t + e^{At} \mathbf{A}^{-1} \mathbf{V}(0) - \mathbf{A}^{-1} \mathbf{V}(0) \tag{18}$$

Since $\mathbf{A} e^{At} = e^{At} \mathbf{A}$ and $\mathbf{A}^{-1} e^{At} = e^{At} \mathbf{A}^{-1}$, it an be verified that $\mathbf{A}^{-1}$ is given by

$$\mathbf{A}^{-1} = \frac{1}{k^2 + f^2} \begin{bmatrix} -k & -f \\ f & -k \end{bmatrix} \tag{IM.19}$$

To simplify the notation, set

$$\mathbf{x}_c = \begin{pmatrix} x_c \\ y_c \end{pmatrix} = -A^{-1} V(0) = \frac{1}{k^2 + f^2} \begin{bmatrix} ku(0) + fv(0) \\ -fu(0) + kv(0) \end{bmatrix} \tag{IM.20}$$

Substituting (IM.20) in (18), we get the exact expression for the position of the water parcel at time t as

$$\mathbf{x}(t) = \mathbf{x}(0) + \mathbf{V}_B t - e^{At} \mathbf{x}_c + \mathbf{x}_c \tag{IM.21}$$

In compound from (IM.21) becomes

$$x(t) = x(0) + V_{1,B} t + x_c - e^{-kt}(x_c \cos ft + y_c \sin ft)$$

$$y(t) = y(0) + V_{2,B} t + y_c - e^{-kt}(y_c \cos ft - x_c \sin ft), \tag{IM.22}$$

where $(x_c, y_c)^T = \mathbf{x}_c$ is given in (IM.20) and $\mathbf{x}(0) = (x(0), y(0))^T$ is the starting position of the chosen water parcel at time $t = 0$.

IM.2 Nondimensional Form of the Solution in (IM.21)

Let $T_0 = 15\,\mathrm{h} = 54 \times 10^3\,\mathrm{s}$ be the time scale and $f_0 = \frac{2\pi}{T_0} = 1.164 \times 10^{-4}\,\mathrm{s}^{-1}$ be the scale for inertial oscillation. Let $L_0 = 1\,\mathrm{km}$ be the spatial scale. Let the velocity scale be $V_0 = L_0 f_0 = 11.64\,\mathrm{cm\,s}^{-1}$. Define the non dimensional quantities using the hat symbol as follows:

$$t = T_0\hat{t}, f = f_0\hat{f}, \ k = f_0\hat{k},$$

$$\{u, v, v_{1,B}, v_{2,B}\} = V_0\{\hat{u}, \hat{v}, \hat{v}_{1,B}, \hat{v}_{2,B}\} \text{ and}$$

$$\{x, y\} = L_0\{\hat{x}, \hat{y}\}.$$

Substituting these in (IM.15) we get the non-dimensional velocity components as

$$\hat{u}(t) = \hat{v}_{1,B} + e^{-2\pi\hat{k}\hat{t}}\left[\hat{u}(0)\cos 2\pi\hat{f}\hat{t} + \hat{v}(0)\sin 2\pi\hat{f}\hat{t}\right] \tag{IM.23}$$

and

$$\hat{v}(t) = \hat{v}_{2,B} + e^{-2\pi\hat{k}\hat{t}}\left[-\hat{u}(0)\sin 2\pi\hat{f}\hat{t} + \hat{v}(0)\cos 2\pi\hat{f}\hat{t}\right] \tag{IM.24}$$

Similarly we get the non dimensional position vector as

$$\hat{x}(t) = \hat{x}(0) + 2\pi\hat{v}_{1,B}\hat{t} - e^{-2\pi\hat{k}\hat{t}}\left[\hat{x}_c\cos 2\pi\hat{f}\hat{t} + \hat{y}_c\sin 2\pi\hat{f}\hat{t}\right] + x_c \tag{IM.25}$$

$$\hat{y}(t) = \hat{y}(0) + 2\pi\hat{v}_{2,B}\hat{t} - e^{-2\pi\hat{k}\hat{t}}\left[-\hat{x}_c\sin 2\pi\hat{f}\hat{t} + \hat{y}_c\cos 2\pi\hat{f}\hat{t}\right] + y_c \tag{IM.26}$$

where

$$\hat{x}_c = \frac{+\hat{k}\hat{u}(0) + \hat{f}\hat{v}(0)}{\hat{k}^2 + \hat{f}^2}$$

and

$$\hat{y}_c = \frac{-\hat{f}\hat{u}(0) + \hat{k}\hat{v}(0)}{\hat{k}^2 + \hat{f}^2}$$

IM.3 True Control

For simplicity, we drop the hat notation. From (IM.23)–(IM.24) it is clear that the position vectors $x(t)$ and $y(t)$ depend on six control variables—$v_{1,B}, v_{2,B}, f, k, u(0)$ and $v(0)$.

Control variables:

$$v_{1,B} = -0.192, v_{2,B} = 0.288,$$

$$u(0) = -1.435, v(0) = -1.435$$

$$f = 0.857 \text{ (period of 17.5 hours)}$$

$$k = 0.015$$

IM.4 Generate Current Observations

In the numerical experiment presented, we assume an idealized state where the observed currents are exact–given by solutions based on true control. Observations are available at $t = 0.500$, 1.25, 2.75, 3.00, and 3.75.

IM.5 Forecast from Erroneous Control

Starting from the erroneous control $v_{1,B} = -0.250, v_{2,B} = 0.250, f = 1.0, k = 0.025, u(0) = -2.0$ and $v(0) = -1.0$, we generate the forecast velocity $(u(t), v(t))^T$ and Lagrangian position $(x(t), y(t))^T$. These forecasts are shown in Fig. IM.2. These plots are called hodographs and are represented by the vector $(u(t), v(t))^T$—$u(t)$ along the horizontal axis (positive towards East) and $v(t)$ along the vertical axis (positive towards North). The Lagrangian trajectory of the water is based on incorrect control. The trajectory extends over a period of $t = 0, 6$ (0–90 h), and is shown in Fig. IM.3.

IM.6 Evolution of Forward Sensitivities

Sensitivities can be found in two ways: directly differentiating $u(t), v(t)$ with respect to $v_{1,B}, v_{2,B}, k, f, u(0), v(0)$ or by deriving the dynamics of evolution of first order sensitivities directly from model equations. Sensitivity of $u(t)$ to $u(0)$ and k is shown in Fig. IM.4 (based on erroneous control).

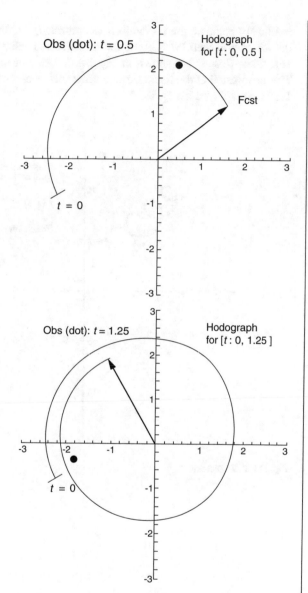

Fig. IM.2 Forecasts at $t = 0.5$ and 1.25 from erroneous control where observations are shown as dots

IM.7 Data Assimilation

Using the observations from Step IM.4 (with the various σ^2) and the forecast made in Step IM.5, compute the forecast errors. Then using the FSM method, iteratively compute the corrections to control. By adding these corrections to the forecast in Step IM.5, compute the corrected value of control and plot the velocities. Results are shown in Fig. IM.5. The adjustment is almost perfect. The unadjusted and adjusted trajectories that overlay the observation at $t =$ 1.25 are displayed in Fig. IM.6.

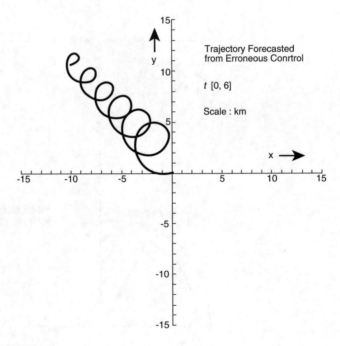

Fig. IM.3 Trajectory

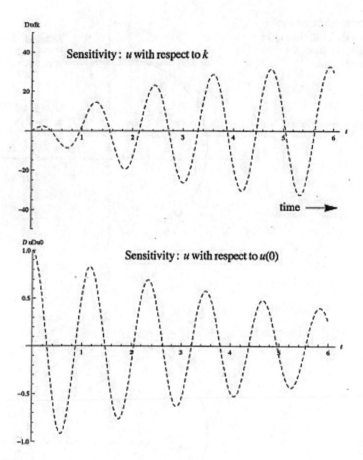

Fig. IM.4 Sensitivity of $u(t)$ to $u(0)$ and k

82

Fig. IM.5 The corrected
value of the control and plot
the velocities using the
corrected control

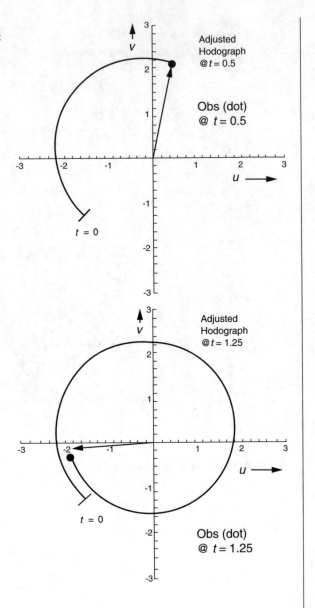

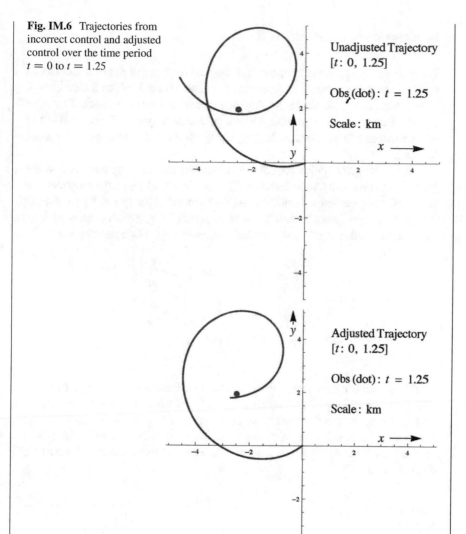

Fig. IM.6 Trajectories from incorrect control and adjusted control over the time period $t = 0$ to $t = 1.25$

Unadjusted Trajectory
[t: 0, 1.25]

Obs (dot): $t = 1.25$

Scale: km

Adjusted Trajectory
[t: 0, 1.25]

Obs (dot): $t = 1.25$

Scale: km

Demonstration: Ocean Inertia II

The governing equation in nondimensional form appear in the problem Ocean Inertia I. In this example, we consider the case where 3 of the 8 elements of true control are made erroneous (primed elements located below). The three elements are: k, u_0, U. The control elements are displayed in Table ASP.1. The true elements are assumed to be slightly different than those used in Ocean Inertia I.

The governing equation for both the trajectory and velocity components are given in the problem Ocean Inertia I. The sensitivity of velocity components u and v with respect to the translation or background velocity (U, V) are simply 0 or 1. The equations governing the sensitivities of u, v with respect to k, u_0 are found by differentiating the momentum equations. For example,

$$\frac{d}{dt}\left(\frac{\partial u}{\partial f}\right) = v \qquad +f\frac{\partial v}{\partial f} - k\frac{\partial u}{\partial f}$$

$$\frac{d}{dt}\left(\frac{\partial v}{\partial k}\right) = -v \qquad -f\frac{\partial u}{\partial k} - k\frac{\partial v}{\partial k}$$

$$\vdots \qquad\qquad\qquad \vdots$$

The solution below stems from solution of the momentum equation and sensitivity equations. The incorrect control must be used since we assume no knowledge of the true control except that it is implicitly contained in the observations that will be created from true control. These observations of u, v are made at times $t_i = 0.4, 1.2, 2.4$ and 3.1. The first order sensitivity matrix is represented as

Table ASP.1 The control elements

True	Error
$f = 1.00$	$f = 1.00$
$k = 0.02$	$k' = 0.025$
$u_0 = 2.00$	$u_0' = 1.75$
$v_0 = 0.00$	$v_0 = 0.00$
$U = 0.30$	$U' = 0.35$
$V = 0.00$	$V = 0.00$
$x_0 = 0.00$	$x_0 = 0.00$
$y_0 = 0.00$	$y_0 = 0.00$

$$
S = \begin{bmatrix}
\left.\dfrac{\partial u}{\partial U}\right|_{t_1} & \left.\dfrac{\partial u}{\partial k}\right|_{t_1} & \left.\dfrac{\partial u}{\partial u_0}\right|_{t_1} \\[2mm]
\left.\dfrac{\partial v}{\partial U}\right|_{t_1} & \left.\dfrac{\partial v}{\partial k}\right|_{t_1} & \left.\dfrac{\partial v}{\partial u_0}\right|_{t_1} \\[2mm]
\left.\dfrac{\partial u}{\partial U}\right|_{t_2} & \left.\dfrac{\partial u}{\partial k}\right|_{t_2} & \left.\dfrac{\partial u}{\partial u_0}\right|_{t_2} \\[2mm]
\left.\dfrac{\partial v}{\partial U}\right|_{t_2} & \left.\dfrac{\partial v}{\partial k}\right|_{t_2} & \left.\dfrac{\partial v}{\partial u_0}\right|_{t_2} \\[2mm]
\left.\dfrac{\partial u}{\partial U}\right|_{t_3} & \left.\dfrac{\partial u}{\partial k}\right|_{t_3} & \left.\dfrac{\partial u}{\partial u_0}\right|_{t_3} \\[2mm]
\left.\dfrac{\partial v}{\partial U}\right|_{t_3} & \left.\dfrac{\partial v}{\partial k}\right|_{t_3} & \left.\dfrac{\partial v}{\partial u_0}\right|_{t_3} \\[2mm]
\left.\dfrac{\partial u}{\partial U}\right|_{t_4} & \left.\dfrac{\partial u}{\partial k}\right|_{t_4} & \left.\dfrac{\partial u}{\partial u_0}\right|_{t_4} \\[2mm]
\left.\dfrac{\partial v}{\partial U}\right|_{t_4} & \left.\dfrac{\partial v}{\partial k}\right|_{t_4} & \left.\dfrac{\partial v}{\partial u_0}\right|_{t_4}
\end{bmatrix}
\begin{matrix}
\leftarrow t_1 = 0.4 \\[6mm]
\leftarrow t_2 = 1.2 \\[6mm]
\leftarrow t_3 = 2.4 \\[6mm]
\leftarrow t_4 = 3.1 \\[6mm]
\end{matrix}
\tag{ASP.1}
$$

The numerical values of elements in this matrix are given below:

$$
S = \begin{bmatrix}
1 & 3.34 & -0.76 \\
0 & 2.43 & -0.55 \\
1 & -3.38 & 0.26 \\
0 & 10.39 & -0.79 \\
1 & 14.64 & -0.55 \\
0 & 10.64 & -040 \\
1 & -16.95 & 0.50 \\
0 & 12.31 & -0.36
\end{bmatrix}
\begin{matrix}
\leftarrow t_1 = 0.4 \\[3mm]
\\
\leftarrow t_2 = 1.2 \\[3mm]
\\
\leftarrow t_3 = 2.4 \\[3mm]
\\
\leftarrow t_4 = 3.1 \\[3mm]
\end{matrix}
\tag{ASP.2}
$$

The difference vector, observations minus forecast follow:

$$
d = \begin{bmatrix}
\tilde{u}(t_1) - u(t_1) \\
\tilde{v}(t_1) - v(t_1) \\
\tilde{u}(t_2) - u(t_2) \\
\tilde{v}(t_2) - v(t_2) \\
\tilde{u}(t_3) - u(t_3) \\
\tilde{v}(t_3) - v(t_3) \\
\tilde{u}(t_4) - u(t_4) \\
\tilde{v}(t_4) - v(t_4)
\end{bmatrix}
=
\begin{bmatrix}
-0.259 \\
-0.152 \\
0.34 \\
-0.257 \\
-0.276 \\
-0.164 \\
0.176 \\
-0.164
\end{bmatrix}
\tag{ASP.3}
$$

The solution is given by

$$
\begin{bmatrix}
\Delta U \\
\Delta k \\
\Delta u_0
\end{bmatrix}
= (S^T S)^{-1} S^T d =
\begin{bmatrix}
-0.050 \\
-0.006 \\
+0.248
\end{bmatrix}
\tag{ASP.4}
$$

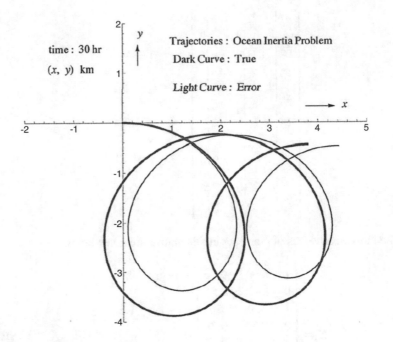

Fig. ASP.1 Trajectories

and the adjusted control is

$$\begin{bmatrix} U \\ k \\ u_0 \end{bmatrix} = \begin{bmatrix} 0.3000 \\ 0.019 \\ 1.998 \end{bmatrix} \qquad \text{(ASP.5)}$$

virtually identical to the corresponding true element of control. The corrected trajectory is nearly identical to the true trajectory. The true and erroneous trajectory are shown in Fig. ASP.1.

Demonstration: Ecosystem

An ecosystem consists of rabbits with an infinite supply of food and foxes that prey on the rabbits. The classic Lotke-Volterra equations describe the dynamics of this predator-prey problem. This dynamical model is based on the following assumptions:

(a) There is ample supply of food for the prey at all times, $t \geq 0$
(b) The food for the predator depends entirely on the prey population

(c) The rate of change of each population is proportional to its size
(d) The predators have limitless appetites

This predator-prey system is governed by a pair of nonlinear first-order ordinary equations given by

$$\frac{dr}{dt} = 2r - \alpha rf$$
$$\frac{df}{dt} = -f + \alpha rf \tag{ES.1}$$

where $r = r(t)$ is the number of rabbits and $f = f(t)$ is the number of foxes at time t and $\alpha \geq 0$ is the interaction parameter. The initial populations are $r(0)$ and $f(0)$. When $\alpha = 0$, there is no interaction between predator and prey and the population of rabbits increases and that of foxes decreases, both exponentially. When $\alpha > 0$, the interaction term allows the fox population to grow at the expense of rabbit population.

Let $\mathbf{x} = (x_1, x_2)^T$ with $x_1 = r$ and $x_2 = f$. Then (ES.1) takes the form (with $\alpha > 0$)

$$\frac{dx_1}{dt} = g_1(\mathbf{x}, \alpha) = 2x_1 - \alpha x_1 x_2$$
$$\frac{dx_2}{dt} = g_2(\mathbf{x}, \alpha) = -x_2 + \alpha x_1 x_2. \tag{ES.2}$$

Setting $\mathbf{g}(\mathbf{x}, \alpha) = (g_1(\mathbf{x}, \alpha), g_2(\mathbf{x}, \alpha))^T$, (ES.2) can be succinctly written as

$$\dot{\mathbf{x}} = \mathbf{g}(\mathbf{x}, \alpha). \tag{ES.3}$$

The set of all vectors $\mathbf{x} \in \mathbb{R}^2$ for which $\mathbf{g}(\mathbf{x}, \alpha) = 0$ is called the system equilibria. It can be verified that there are two equilibrium states for (ES.3) given by

$$\mathbf{E}_1 = (0, 0)^T \text{ and } \mathbf{E}_2 = \frac{1}{\alpha}(1, 2)^T. \tag{ES.4}$$

The Jacobian of $\mathbf{g}$ is given by

$$\mathbf{D_x}(\mathbf{g}) = \begin{bmatrix} \frac{\partial g_1}{\partial x_1} & \frac{\partial g_1}{\partial x_2} \\ \frac{\partial g_2}{\partial x_1} & \frac{\partial g_2}{\partial x_2} \end{bmatrix} = \begin{bmatrix} 2 - \alpha x_2 & -\alpha x_1 \\ \alpha x_2 & -1 + \alpha x_1 \end{bmatrix} \tag{ES.5}$$

At the equilibrium $\mathbf{E}_1$,

$$\mathbf{D_x(g)}|_{\mathbf{x}=(0,0)^T} \begin{bmatrix} 2 & 0 \\ 0 & -1 \end{bmatrix} \tag{ES.6}$$

whose eigenvalues are $\lambda_1 = 2$ and $\lambda_2 = -1$.

Hence, in a small neighborhood around the origin with $x_1 > 0$ and $x_2 > 0$, we readily see that $x_1(t)$ grows as $e^{2t} x_1(0)$ and $x_2(t)$ decreases as $e^{-t} x_2(0)$. As the trajectories move away from $\mathbf{E}_1$, the interaction terms become effective and the local analysis given above no longer holds.

Likewise, at the equilibrium $\mathbf{E}_2$

$$\mathbf{D_x(g)}|_{x=\frac{1}{\alpha}(1,2)^T} = \begin{bmatrix} 0 & -1 \\ 2 & 0 \end{bmatrix} \tag{ES.7}$$

whose eigenvalues are purely imaginary with $\lambda_{1,2} = \pm i \sqrt{2}$. Consequently, if we start in a small neighborhood around $\mathbf{E}_2$, the solution will oscillate around E_2.

Generation of Observations

It is assumed that the populations are observed at different times. For rabbits, the population is observed at times $t = 1.5$ and 6.0 while the fox population is observed at $t = 2.0$ and 8.0. It is further assumed that true $\alpha = 0.05$ and erroneous $\alpha = 0.06$.

Observations of rabbits z_r and foxes z_f are found by solving (ES.2) with true $\alpha = 0.06$ and true initial conditions $x_1(0) = 30$ and $x_2(0) = 15$. The forecasts are made by solving (ES.2) with true initial conditions but erroneous $\alpha = 0.05$. Forecasts and observations are recorded in Table ES.1. A plot of the forecast overlaying the observations is shown in Fig. ES.1, Interestingly, there is a lag in the maximum fox population relative to the maximum in the rabbit population (Explain based on dynamics).

Table ES.1 Observation of two populations

		Observations		Forecast	
Index i	Time t_i	z_r	z_f	$x_1(t)$	$x_2(t)$
1	1.5	20	–	12	–
2	2.0	–	71	–	52
3	6.0	45	–	25	–
4	8.0	–	31	–	22

Forecast Error Correction Using FSM

We seek to correct the forecast errors by adjusting the model parameters using FSM. Let $\mathbf{v}_1(t) = \left(\frac{\partial x_1}{\partial \alpha}, \frac{\partial x_2}{\partial \alpha} \right)^T$ It can be verified that

$$\dot{\mathbf{v}}_1(t) = \mathbf{D}_{\mathbf{x}(t)}(\mathbf{g})\mathbf{v}_1(t) + \mathbf{D}_\alpha(\mathbf{g}) \qquad \text{(ES.8)}$$

where the Jacobian $\mathbf{D}_{\mathbf{x}(t)}(\mathbf{g})$ is given in (ES.5) and the Jacobian

$$\mathbf{D}_\alpha = \begin{bmatrix} -x_1 & x_2 \\ x_1 & x_2 \end{bmatrix} \qquad \text{(ES.9)}$$

with $\mathbf{v}_1(0) = 0$.

The evolution of the components of $\mathbf{v}(t)[\mathbf{v}_1(t)]$ as a function of time by simultaneously solving (ES.2) and (ES.8). Results are shown in Figs. ES.2 and ES.3.

Data Assimilation Using FSM

In this example, the set of equations governing the problem follow:

$$\frac{dr}{dt} = 2r - \alpha rf$$

$$\frac{df}{dt} = -f + \alpha fr$$

$$\frac{d}{dt}\left(\frac{df}{d\alpha}\right) = 2\frac{\partial r}{\partial \alpha} - rf - \alpha r\frac{\partial f}{\partial \alpha} - \alpha f\frac{\partial r}{\partial \alpha}$$

$$\frac{d}{dt}\left(\frac{df}{d\alpha}\right) = -\frac{\partial f}{\partial \alpha} + rf + \alpha\frac{\partial r}{\partial f}\cdot f + \alpha r\frac{\partial f}{\partial \alpha} \qquad \text{(ES.10)}$$

given $r(0) = 30, f(0) = 15, \frac{\partial r}{\partial \alpha}(0) = 0, \frac{\partial f}{\partial \alpha}(0) = 0$

The first-order FSM solution subject to these constraints yields $\alpha = 0.46$ after one iteration. This result is graphed in Fig. ES.4. A second iteration yields $\alpha = 0.49$, very close to the exact solution.

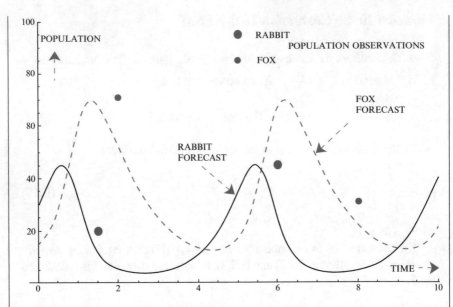

Fig. ES.1 Populations of foxes and rabbits

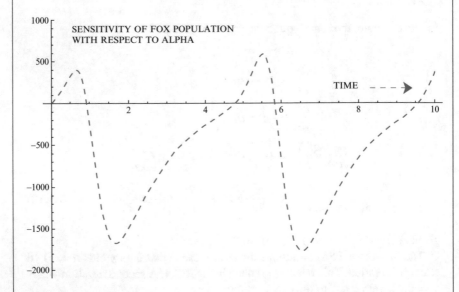

Fig. ES.2 Sensitivity of fox population w.r.t. α

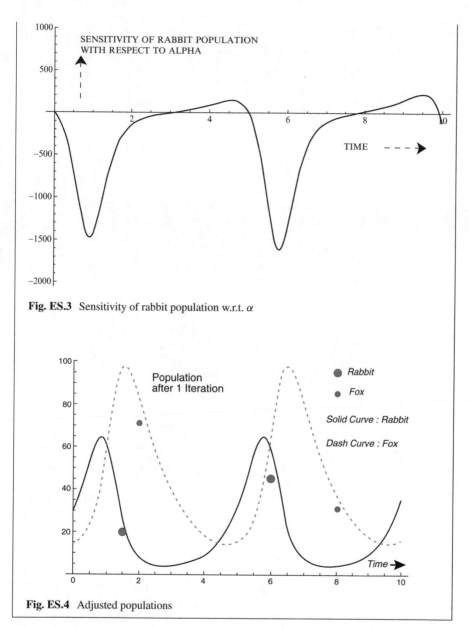

Fig. ES.3 Sensitivity of rabbit population w.r.t. α

Fig. ES.4 Adjusted populations

3.5   Notes and References

This Chapter closely follows Lakshmivarahan and Lewis (2010). For a detailed treatment of 4D-VAR methods refer to Lewis et al. (2006, Chaps. 22–25).

Chapter 4
Forward Sensitivity Method: General Case

In this chapter we generalize the results of Chaps. 2 and 3 to nonlinear dynamical systems. In Sect. 4.1, the dynamics of evolution of the first order sensitivities are given. The intrinsic relation between the adjoint and forward sensitivities are explored in Sect. 4.2. In Sect. 4.3 forward sensitivity based data assimilation scheme is derived. Exercises and Notes and References are contained in Sects. 4.4 and 4.5, respectively.

4.1 Dynamics of Evolution of First Order Forward Sensitivities

Let $\mathbf{x} \in \mathbb{R}^n$ denote the state of a discrete time nonlinear dynamical system and let $\boldsymbol{\alpha} \in \mathbb{R}^p$ be the parameter vector that includes both the physical parameters and boundary conditions. Let the one step state transition map $\mathbf{M} : \mathbb{R}^n \times \mathbb{R}^p \to \mathbb{R}^n$ be the vector field that defines the model dynamics, where $\mathbf{M}(\mathbf{x}, \boldsymbol{\alpha}) = (\mathbf{M}_1(\mathbf{x}, \boldsymbol{\alpha}), \mathbf{M}_2(\mathbf{x}, \boldsymbol{\alpha}), \dots \mathbf{M}_n(\mathbf{x}, \boldsymbol{\alpha}))^T$. The discrete time model dynamics is given by

$$\mathbf{x}(k+1) = \mathbf{M}(\mathbf{x}(k), \boldsymbol{\alpha})) \tag{4.1.1}$$

where $\mathbf{x}(0)$ is the initial condition. It follows that the solution $\mathbf{x}(k) = \mathbf{x}(k, \mathbf{x}(0), \boldsymbol{\alpha})$ depends continuously on $\mathbf{x}(0)$ and $\boldsymbol{\alpha}$ and can be altered by changing $\mathbf{x}(0)$ and/or $\boldsymbol{\alpha}$. Let

$$\mathbf{u}_1(k) = \frac{\partial \mathbf{x}(k)}{\partial \mathbf{x}(0)} = \left[\frac{\partial \mathbf{x}_i(k)}{\partial \mathbf{x}_j(0)} \right] \in \mathbb{R}^{n \times n}, \tag{4.1.2}$$

© Springer International Publishing Switzerland 2017
S. Lakshmivarahan et al., *Forecast Error Correction using Dynamic Data Assimilation*, Springer Atmospheric Sciences, DOI 10.1007/978-3-319-39997-3_4

denote the first order forward sensitivity matrix whose (i, j)th element $\frac{\partial \mathbf{x}_i(k)}{\partial \mathbf{x}_j(0)}$ is the sensitivity of the ith component $\mathbf{x}_i(k)$ of $\mathbf{x}(k)$ at time k with respect to the jth component $\mathbf{x}_j(0)$ of the initial condition $\mathbf{x}(0)$. Similarly, define

$$\mathbf{v}_1(k) = \frac{\partial \mathbf{x}(k)}{\partial \boldsymbol{\alpha}} = \left[\frac{\partial \mathbf{x}_i(k)}{\partial \alpha_j} \right] \in \mathbb{R}^{n \times p}, \tag{4.1.3}$$

denote the forward sensitivity matrix whose (i, j)th element $\frac{\partial \mathbf{x}_i(k)}{\partial \alpha_j}$ is the sensitivity of the ith component $\mathbf{x}_i(k)$ of $\mathbf{x}(k)$ at time k with respect to the jth component α_j of the parameter $\boldsymbol{\alpha}$. Our goal in this section is to define the dynamics of evolution of sensitivity matrices $\mathbf{u}_1(k)$ and $\mathbf{v}_1(k)$ defined above.

4.1.1   Dynamics of Evolution of $\mathbf{u}_1(k)$

Consider the ith component of the model dynamics in (4.1.1) given by

$$\mathbf{x}_i(k + 1) = \mathbf{M}_i(\mathbf{x}(k), \boldsymbol{\alpha}). \tag{4.1.4}$$

Differentiating both sides with respect to $\mathbf{x}_j(0)$, we get

$$\frac{\partial \mathbf{x}_i(k + 1)}{\partial \mathbf{x}_j(0)} = \sum_{r=1}^{n} \frac{\partial \mathbf{M}_i(\mathbf{x}(k), \boldsymbol{\alpha})}{\partial \mathbf{x}_r(k)} \frac{\partial \mathbf{x}_r(k)}{\partial \mathbf{x}_j(0)}, \tag{4.1.5}$$

with the initial condition given by

$$\left. \frac{\partial \mathbf{x}_i(k)}{\partial \mathbf{x}_j(0)} \right|_{k=0} = \begin{cases} 1 \text{ if } i = j \\ 0 \text{ if } i \neq j \end{cases}. \tag{4.1.6}$$

The sum on the right hand side of (4.1.5) is the inner product of the ith row of the Jacobian of the model map $\mathbf{M}$ with respect to the state vector $\mathbf{x}$ given by

$$\mathbf{D}_{\mathbf{x}(k)}(\mathbf{M}) = \frac{\partial \mathbf{M}(\mathbf{x}(k), \boldsymbol{\alpha})}{\partial \mathbf{x}(k)} = \left[\frac{\partial \mathbf{M}_i(\mathbf{x}(k), \boldsymbol{\alpha})}{\partial \mathbf{x}_r(k)} \right] \in \mathbb{R}^{n \times n} \tag{4.1.7}$$

and the jth column of the forward sensitivity matrix in (4.1.2). The n^2 equations in (4.1.5)–(4.1.7) can be succinctly written as

$$\mathbf{u}_1(k + 1) = \mathbf{D}_{\mathbf{x}(k)}(\mathbf{M}) \, \mathbf{u}_1(k) \tag{4.1.8}$$

with

$$\mathbf{u}_1(0) = \mathbf{I}$$

Iterating (4.1.8), it is immediate that

$$\mathbf{u}_1(k) = \mathbf{D}_{\mathbf{x}(k-1)}(\mathbf{M})\,\mathbf{D}_{\mathbf{x}(k-2)}(\mathbf{M})\dots\mathbf{D}_{\mathbf{x}(0)}(\mathbf{M}) = \prod_{i=0}^{k-1}{}' \mathbf{D}_{\mathbf{x}(i)}(\mathbf{M}) \qquad (4.1.9)$$

where $\prod'$ denotes the product of Jacobian in decreasing time index from time $(k-1)$ to 0.

4.1.2 Dynamics of Evolution of $\mathbf{v}_1(k)$

Differentiating (4.1.4) with respect to α_j, we get

$$\frac{\partial x_i(k+1)}{\partial \alpha_j} = \sum_{r=1}^{n} \frac{\partial M_i(\mathbf{x}(k), \boldsymbol{\alpha})}{\partial x_r(k)} \frac{\partial x_r(k)}{\partial \alpha_j} + \frac{\partial M_i(\mathbf{x}(k), \boldsymbol{\alpha})}{\partial \alpha_j}, \qquad (4.1.10)$$

with the initial condition

$$\frac{\partial x_i(0)}{\partial \alpha_j} = 0 \text{ for all } 1 \le i \le n, \text{ and } 1 \le j \le p. \qquad (4.1.11)$$

The first term on the right hand side of (4.1.10) arises through the implicit dependence of $\mathbf{M}(\mathbf{x}, \boldsymbol{\alpha})$ on α_j through the state $\mathbf{x}(k)$ and the second term is due to the explicit dependence of $\mathbf{M}(\mathbf{x}, \boldsymbol{\alpha})$ on $\boldsymbol{\alpha}$. Clearly, the first term is the inner product of the ith row of $\mathbf{D}_{\mathbf{x}(k)}(\mathbf{M})$ in (4.1.7) and the jth column of the sensitivity matrix in (4.1.3). The second term in (4.1.10) is the (i, j)th entry of the model Jacobian

$$\mathbf{D}_\alpha = \frac{\partial \mathbf{M}}{\partial \boldsymbol{\alpha}} = \left[\frac{\partial M_i(\mathbf{x}(k), \boldsymbol{\alpha})}{\partial \alpha_j} \right] \in \mathbb{R}^{n \times p}, \qquad (4.1.12)$$

of the model map $\mathbf{M}$ with respect to the parameter vector $\boldsymbol{\alpha}$. The np equations in (4.1.10)–(4.1.11) can be succinctly written as

$$\mathbf{v}_1(k+1) = \mathbf{D}_{\mathbf{x}(k)}(\mathbf{M})\,\mathbf{v}_1(k) + \mathbf{D}_\alpha(\mathbf{M}), \qquad (4.1.13)$$

with the initial condition $\mathbf{v}_1(0) = 0$. Iterating (4.1.13), it can be verified that

$$\mathbf{v}_1(k) = \sum_{i=0}^{k-1} \left[\prod_{q=i+1}^{k-1}{}' \mathbf{A}(q) \right] \mathbf{B}(i) \qquad (4.1.14)$$

where $\mathbf{A}(k) = \mathbf{D}_{\mathbf{x}(k)}\mathbf{M}(\mathbf{x}(k), \boldsymbol{\alpha})$ and $\mathbf{B}(k) = \mathbf{D}_\alpha \mathbf{M}(\mathbf{x}(k), \boldsymbol{\alpha})$ for simplicity in notation. (Refer to Problem 2.3 in Chap. 2.)

4.1.3 Propagation of Perturbation and Forward Sensitivity

Recall that the forecast trajectory of the model (4.1.1) depends continuously on the initial condition $\mathbf{x}(0) \in \mathbb{R}^n$ and the parameter $\boldsymbol{\alpha} \in \mathbb{R}^p$. To emphasize this dependence, the initial condition-parameter pair is called the control variables in data assimilation literature.

Let $\delta\mathbf{c} = \left(\delta\mathbf{x}^T(0), \delta\boldsymbol{\alpha}^T\right)^T$ be a perturbation in the control $\overline{\mathbf{c}}$ arising from the perturbation $\delta\mathbf{x}(0)$ in $\overline{\mathbf{x}}(0)$ and $\delta\boldsymbol{\alpha}$ in $\overline{\boldsymbol{\alpha}}$. Let $\mathbf{x}(k)$ be the new forecast trajectory of the model starting from the new control $\mathbf{c} = \overline{\mathbf{c}} + \delta\mathbf{c}$. Our goal in this section is to derive the evolution of the first-order perturbation $\delta\mathbf{x}(k)$ in $\overline{\mathbf{x}}(k)$ induced by the perturbation $\delta\mathbf{c}$ in $\overline{\mathbf{c}}$ and derive its dependence on the forward sensitivity matrices $\mathbf{u}_1(k)$ and $\mathbf{v}_1(k)$ introduced in the previous two subsections.

Recall that the base forecast is given by

$$\overline{\mathbf{x}}(k+1) = \mathbf{M}\left(\overline{\mathbf{x}}(k), \overline{\boldsymbol{\alpha}}\right) \tag{4.1.15}$$

with $\overline{\mathbf{c}}$ as the control and the perturbed forecast given by

$$\mathbf{x}(k+1) = \mathbf{M}\left(\mathbf{x}(k), \boldsymbol{\alpha}\right), \tag{4.1.16}$$

with $\mathbf{c} = \overline{\mathbf{c}} + \delta\mathbf{c}$ as the control. Let

$$\delta\mathbf{x}(k) = \mathbf{x}(k) - \overline{\mathbf{x}}(k), \tag{4.1.17}$$

be the first-order approximation to the difference between the perturbed and the base forecast trajectories. Then

$$\begin{aligned}
\delta\mathbf{x}(k+1) &= \mathbf{x}(k+1) - \overline{\mathbf{x}}(k+1) \\
&= \mathbf{M}\left(\mathbf{x}(k), \boldsymbol{\alpha}\right) - \mathbf{M}\left(\overline{\mathbf{x}}(k), \overline{\boldsymbol{\alpha}}\right) \\
&= \mathbf{M}\left(\overline{\mathbf{x}}(k) + \delta\mathbf{x}(k), \overline{\boldsymbol{\alpha}} + \delta\boldsymbol{\alpha}\right) - \mathbf{M}\left(\overline{\mathbf{x}}(k), \overline{\boldsymbol{\alpha}}\right) \\
&= \mathbf{A}(k)\delta\mathbf{x}(k) + \mathbf{B}(k)\delta\boldsymbol{\alpha}, \tag{4.1.18}
\end{aligned}$$

where the last line is obtained by applying the first-order Taylor series (Appendix A) to the expression in the previous line and

$$\mathbf{A}(k) = \mathbf{D}_{\overline{\mathbf{x}}(k)}\left(M\right) \in \mathbb{R}^{n\times n}, \text{ and } \mathbf{B}(k) = D_{\overline{\boldsymbol{\alpha}}}\left(\mathbf{M}\right) \in \mathbb{R}^{n\times p}. \tag{4.1.19}$$

are the Jacobians of $\mathbf{M}$ with respect to $\mathbf{x}$ and $\boldsymbol{\alpha}$, respectively.

Equation (4.1.18) governs the evolution of the perturbation which is a linear, non-autonomous and non-homogeneous first-order difference equation.

By exploiting the structural similarity of (4.1.18) with the recurrence in Problem 2.3 in Chap. 2, it readily follows (left as an exercise—see Problem 4.1) that

$$\delta\mathbf{x}(k) = \mathbf{u}_1(k)\delta\mathbf{x}(0) + \mathbf{v}_1(k)\delta\boldsymbol{\alpha}, \tag{4.1.20}$$

where

$$\mathbf{u}_1(k) = \left[\frac{\partial \overline{\mathbf{x}}(k)}{\partial \overline{\mathbf{x}}(0)} \right] = \prod_{i=0}^{k-1}{}' \mathbf{A}(i) \tag{4.1.21}$$

and

$$\mathbf{v}_1(k) = \left[\frac{\partial \overline{\mathbf{x}}(k)}{\partial \overline{\alpha}} \right] = \sum_{i=0}^{k-1} \left[\prod_{q=i+1}^{k-1}{}' \mathbf{A}(q) \right] \mathbf{B}(i) \delta \alpha \tag{4.1.22}$$

are the forward sensitivities of $\overline{\mathbf{x}}(k)$ with respect to $\overline{\mathbf{x}}(0)$ and $\overline{\alpha}$, respectively. The evolution equation (4.1.20) plays a critical role in data assimilation using FSM.

4.2 On the Relation Between Adjoint and Forward Sensitivities

Let $\mathbf{z} \in \mathbb{R}^m$ be the observation vector and $\mathbf{h} : \mathbb{R}^n \to \mathbb{R}^m$ be the (nonlinear) map (also known as the forward operator) that defines the relation between the model state $\mathbf{x} \in \mathbb{R}^n$ and the observation $\mathbf{z} \in \mathbb{R}^m$, where

$$\mathbf{z} = \mathbf{h}(\mathbf{x}) + \mathbf{v}. \tag{4.2.1}$$

$\mathbf{v} \in \mathbb{R}^m$ represents the unavoidable and unobservable observation random noise vector which is assumed to have a multivariate normal distribution with mean zero and known covariance matrix $\mathbf{R}$, that is, $\mathbf{v} \sim N(0, \mathbf{R})$ where $\mathbf{R}$ is symmetric and positive definite.

It is assumed that we are given N observations, $\mathbf{z}(k_i), 1 \leq i \leq N$, where the observation times are ordered as follows:

$$0 < k_1 < k_2 < \ldots < k_N. \tag{4.2.2}$$

Let $\overline{\mathbf{c}} = (\overline{\mathbf{x}}(0), \overline{\alpha})$ be the control from which we generate the forecast using the model (4.1.1). Let $\overline{\mathbf{x}}(k) : k \geq 0$ be the resulting forecast trajectory. Our basic assumption in this chapter is that (4.1.1) is a "good" model in the sense it is "faithful" to the physical phenomenon it is expected to capture. Hence, the forecast error, if any, as measured by

$$\overline{\mathbf{f}}(k) = \mathbf{z}(k) - \mathbf{h}(\overline{\mathbf{x}}(k)) \tag{4.2.3}$$

is primarily due to the incorrect control, $\overline{\mathbf{c}}$ used to generate the forecast, $\overline{\mathbf{x}}(k)$.

Determination of the correction $\delta\mathbf{c}$ to $\bar{\mathbf{c}}$ that will deliver an improved forecast $\mathbf{x}(k)$ starting from the new control $c = \bar{\mathbf{c}} + \delta\mathbf{c}$ becomes the central problem. In short, we strive to ensure that resulting forecast error

$$\mathbf{f}(k) = \mathbf{z}(k) - \mathbf{h}\,(\mathbf{x}(k)) \tag{4.2.4}$$

is essentially random noise.

To this end, we consider a functional $J(\bar{\mathbf{c}})$ which is the sum of the squares of the energy norm of the forecast error in (4.2.3) as

$$J(\bar{\mathbf{c}}) = \frac{1}{2}\sum_{i=1}^{N}\bar{\mathbf{f}}^{T}(k_i)\mathbf{R}_i^{-1}\mathbf{f}(k_i), \tag{4.2.5}$$

where $\mathbf{R}_i$ is the covariance of the noise vector $\mathbf{v}(k_i)$ that affects the observation $\mathbf{z}(k_i)$ for $1 \le i \le N$.

Let δJ be the first variation in $J(\bar{\mathbf{c}})$ induced by the first variation $\delta\mathbf{c}$ in $\bar{\mathbf{c}}$. Then

$$\delta J = J(\bar{\mathbf{c}} + \delta\mathbf{c}) - J(\bar{\mathbf{c}}). \tag{4.2.6}$$

From first principles, we also know that

$$\delta J = \langle \nabla_{\bar{\mathbf{c}}}J(\bar{\mathbf{c}}), \delta\mathbf{c} \rangle, \tag{4.2.7}$$

where $\nabla_{\bar{\mathbf{c}}}J(\bar{\mathbf{c}})$ is called adjoint sensitivity vector. Thus, we can get an explicit expression for the first variation of $J(\bar{\mathbf{c}})$ by computing the gradient of each of the terms in the sum on the right hand side of (4.2.5) with respect to $\bar{\mathbf{c}}$. To this end, consider a typical term given by

$$Q(\bar{\mathbf{x}}) = \frac{1}{2}\bar{\mathbf{f}}^{T}\mathbf{R}^{-1}\mathbf{f} = \frac{1}{2}\left[\mathbf{z} - \mathbf{h}(\bar{\mathbf{x}})\right]^{T}\mathbf{R}^{-1}\left[\mathbf{z} - h(\bar{\mathbf{x}})\right]$$
$$= \frac{1}{2}\underbrace{\mathbf{z}^{T}\mathbf{R}^{-1}\mathbf{z}}_{I} - \underbrace{\mathbf{z}^{T}\mathbf{R}^{-1}\mathbf{h}(\bar{\mathbf{x}})}_{II} + \frac{1}{2}\underbrace{\mathbf{h}^{T}(\bar{\mathbf{x}})\mathbf{R}^{-1}\mathbf{h}(\bar{\mathbf{x}})}_{III}. \tag{4.2.8}$$

The gradient of term I with respect to $\mathbf{x}$ is zero, and accordingly it does not contribute to the first variation, δQ of $Q(x)$ in (4.2.8) set to zero. Term II can be rewritten as

$$II = \mathbf{z}^{T}\mathbf{R}^{-1}\mathbf{h}(\bar{\mathbf{x}}) = \mathbf{a}^{T}\mathbf{h}(\bar{\mathbf{x}})$$

with $\mathbf{a} = \mathbf{R}^{-1}\mathbf{z}$. From Appendix A, it follows that

$$\nabla_{\mathbf{x}}\left(\mathbf{z}^{T}\mathbf{R}^{-1}\mathbf{h}(\bar{\mathbf{x}})\right) = \mathbf{D}_{\bar{\mathbf{x}}}^{T}(h)\mathbf{R}^{-1}\mathbf{z} \tag{4.2.9}$$

Similarly, the gradient of term *III* is given by (refer to Appendix A)

$$\nabla_{\bar{\mathbf{x}}}\left[\mathbf{h}^T(\bar{\mathbf{x}})\mathbf{R}^{-1}\mathbf{h}(\bar{\mathbf{x}})\right] = 2\mathbf{D}_{\bar{\mathbf{x}}}^T(\mathbf{h})\mathbf{R}^{-1}\mathbf{h}(\bar{\mathbf{x}}). \tag{4.2.10}$$

Combining (4.2.8)–(4.2.10), we get

$$\delta Q(\bar{\mathbf{x}}) = \langle \nabla_{\bar{x}}Q(\bar{\mathbf{x}}), \delta\mathbf{x} \rangle, \tag{4.2.11}$$

where

$$\nabla_{\bar{x}}Q(\bar{\mathbf{x}}) = -\mathbf{D}_{\bar{x}}^T(\mathbf{h})\mathbf{R}^{-1}\bar{f} = -\mathbf{D}_{\bar{x}}^T(\mathbf{h})\bar{\mathbf{e}} = -\bar{\boldsymbol{\eta}}, \tag{4.2.12}$$

$\bar{\mathbf{f}} = (\mathbf{z} - \mathbf{h}(\bar{\mathbf{x}})) \in \mathbb{R}^m$ is the forecast error in the observation space, and this error in normalized from is given by $\bar{\mathbf{e}} = \mathbf{R}^{-1}\bar{\mathbf{f}} \in \mathbb{R}^n$, and $\bar{\boldsymbol{\eta}} \in \mathbb{R}^n$ is the forecast error viewed from the model space $\mathbb{R}^n$.

However, $\delta\mathbf{x}$ is related to $\delta\mathbf{x}(0)$ and $\delta\boldsymbol{\alpha}$ through the forward sensitivity functions $\mathbf{u}_1$ and $\mathbf{v}_1$ as given in (4.1.20). Substituting (4.1.20) in (4.2.11) and using the adjoint property, we get ($\mathbf{u}_1 \in \mathbb{R}^{n \times n}$ and $\mathbf{v}_1 \in \mathbb{R}^{n \times p}$)

$$\begin{aligned}
\delta Q(\bar{\mathbf{x}}) &= - \langle \bar{\boldsymbol{\eta}}, \mathbf{u}_1\delta\mathbf{x}(0) + \mathbf{v}_1\delta\boldsymbol{\alpha} \rangle \\
&= - \langle \mathbf{u}_1^T\bar{\boldsymbol{\eta}}, \delta\mathbf{x}(0) \rangle - \langle \mathbf{v}_1^T\bar{\boldsymbol{\eta}}, \delta\boldsymbol{\alpha} \rangle \\
&= - \langle \nabla_{\bar{x}}Q(\bar{\mathbf{x}}), \delta\mathbf{x}(0) \rangle - \langle \nabla_{\bar{\alpha}}Q(\bar{\mathbf{x}}), \delta\boldsymbol{\alpha} \rangle
\end{aligned} \tag{4.2.13}$$

with

$$\nabla Q(\bar{\mathbf{x}}) = -\mathbf{u}_1^T\bar{\boldsymbol{\eta}} \quad \text{and} \quad \nabla Q(\bar{\boldsymbol{\alpha}}) = -\mathbf{v}_1^T\bar{\boldsymbol{\eta}}. \tag{4.2.14}$$

Hence

$$\begin{aligned}
\nabla Q(\bar{\mathbf{c}}) &= \left([\nabla_{\bar{x}}Q(\bar{\mathbf{x}})]^T, [\nabla_{\bar{\alpha}}Q(\bar{\mathbf{x}})]^T\right)^T \\
&= -\left((\mathbf{u}_1^T\bar{\boldsymbol{\eta}})^T, (\mathbf{v}_1^T\bar{\boldsymbol{\eta}})^T\right)^T \in \mathbb{R}^{n+p},
\end{aligned} \tag{4.2.15}$$

and thus the gradient of $Q(\mathbf{x})$ with respect to the control $\bar{\mathbf{c}}$ is the concatenation of the column vector $\mathbf{u}_1^T\bar{\boldsymbol{\eta}} \in \mathbb{R}^n$ and the column vector $\mathbf{v}_1^T\bar{\boldsymbol{\eta}} \in \mathbb{R}^p$. Since the gradient of the sum is the sum of the gradient, combining (4.2.5), (4.2.8) and (4.2.15), it readily follows that

$$\nabla_{\bar{c}}J(\bar{\mathbf{c}}) = -\sum_{i=1}^{N}\left[\left(\mathbf{u}_1^T(k_i)\bar{\boldsymbol{\eta}}(k_i)\right)^T, \left(\mathbf{v}_1^T(k_i)\bar{\boldsymbol{\eta}}(k_i)\right)^T\right]^T \in \mathbb{R}^{n+p}. \tag{4.2.16}$$

Substituting (4.2.16) in (4.2.7), we get the expression for the first variation δJ of $J(\bar{\mathbf{c}})$.

Since the first variation is zero at the minimum, the necessary condition for the minimum of $J(\overline{\mathbf{c}})$ translates into the vanishing of $\nabla_{\overline{\mathbf{c}}} J(\overline{\mathbf{c}})$ in (4.2.16). Besides characterizing the necessary condition for the minima of $J(\overline{\mathbf{c}})$, $\nabla_{\overline{\mathbf{c}}} J(\overline{\mathbf{c}})$ provides information on the fine structure of the adjoint sensitivity. Thus

$$\nabla_{\overline{\mathbf{x}}(0)} J(\overline{\mathbf{c}}) = -\sum_{i=1}^{N} \mathbf{u}_1^T(k_i)\, \overline{\boldsymbol{\eta}}(k_i)$$

$$= -\sum_{i=1}^{N} \mathbf{u}_1^T(k_i)\, \mathbf{D}_x^T(\mathbf{h}(\overline{\mathbf{x}}(k_i)))\, \mathbf{R}^{-1}[\mathbf{z}(k_i) - \mathbf{h}(\overline{\mathbf{x}}(k_i))], \quad (4.2.17)$$

which is the sum of the product of the transpose of the forward sensitivity matrix $\mathbf{u}_1(k)$ and the transformed weighted forecast error $\overline{\boldsymbol{\eta}}(k_i)$ viewed from the model space. Similarly,

$$\nabla_{\overline{\boldsymbol{\alpha}}} J(\overline{\mathbf{c}}) = -\sum_{i=1}^{N} \mathbf{v}_1^T(k_i)\, \overline{\boldsymbol{\eta}}(k_i)$$

$$= -\sum_{i=1}^{N} \mathbf{v}_1^T(k_i)\, \mathbf{D}_x^T(\mathbf{h}(\overline{\mathbf{x}}(k_i)))\, \mathbf{R}^{-1}[\mathbf{z}(k_i) - \mathbf{h}(\overline{\mathbf{x}}(k_i))] \quad (4.2.18)$$

has a similar interpretation.

4.3 Data Assimilation Using FSM

To bring out the key features of our approach, we consider two cases as we did in Chap. 2.

4.3.1 Case 1 Single Observation

Let $\mathbf{z}(k) \in \mathbb{R}^m$ be the single vector observation available at time k. Then the forecast error is given by

$$\overline{\mathbf{f}}(k) = \mathbf{z}(k) - \mathbf{h}(\overline{\mathbf{x}}(k)). \qquad (4.3.1)$$

Since the forecast $\overline{\mathbf{x}}(k)$ is generated by the model (4.1.1) starting from a guess control $\overline{\mathbf{c}} = (\overline{\mathbf{x}}^T(0), \boldsymbol{\alpha}^T)^T$, it is reasonable to expect that the forecast error $\overline{\mathbf{f}}(k)$ in (4.3.1) may not be close to zero.

Our goal is to find the correction $\delta\mathbf{c} = \left(\delta\mathbf{x}^T(0), \delta\boldsymbol{\alpha}^T\right)^T$ to $\bar{\mathbf{c}}$ such that the new forecast, $\mathbf{x}(k)$ starting from $\mathbf{c} = (\bar{\mathbf{c}} + \delta\mathbf{c})$ delivers a difference

$$\mathbf{f}(k) = \mathbf{z}(k) - \mathbf{h}\left(\mathbf{x}(k)\right) \tag{4.3.2}$$

that is purely random—that is, free from a deterministic error.

Recall that $\delta\mathbf{x}(k)$ is the perturbation of the state $\bar{\mathbf{x}}(k)$ induced by the initial perturbation $\delta\mathbf{c}$ in the control vector $\bar{\mathbf{c}}$, where $\delta\bar{\mathbf{x}}(0)$ and $\delta\bar{\boldsymbol{\alpha}}$ are functions of the forward sensitivity matrices $\mathbf{u}_1$ and $\mathbf{v}_1$. This $\delta\mathbf{x}(k)$ induces a first variation in $\mathbf{h}\left(\bar{\mathbf{x}}\right)$ give by

$$\delta\mathbf{h} = \mathbf{h}\left(\mathbf{x}(k)\right) - \mathbf{h}\left(\bar{\mathbf{x}}(k)\right) = \mathbf{D}_{\bar{\mathbf{x}}}(h)\delta\mathbf{x}(k), \tag{4.3.3}$$

where $\mathbf{D}_{\mathbf{x}}(\mathbf{h}) \in \mathbb{R}^{m\times n}$ is the Jacobian of $\mathbf{h}(\mathbf{x})$. Substituting (4.3.3) in (4.3.2), we get

$$\begin{aligned}
\mathbf{f}(k) &= \mathbf{z}(k) - \mathbf{h}\left(\mathbf{x}(k)\right) \\
&= \mathbf{z}(k) - \mathbf{h}\left(\bar{\mathbf{x}}(k)\right) - \delta\mathbf{h} \\
&= \bar{\mathbf{f}}(k) - \delta\mathbf{h} \\
&= \bar{\mathbf{f}}(k) - \mathbf{D}_{\bar{\mathbf{x}}}(\mathbf{h})\delta\mathbf{x}(k). \tag{4.3.4}
\end{aligned}$$

Thus, a condition for the annihilation of the deterministic part of the forecast error $\mathbf{f}(k)$ is that $\delta\mathbf{x}(k)$ satisfy the following relation:

$$\mathbf{D}_{\bar{\mathbf{x}}}(\mathbf{h})\delta\mathbf{x}(k) = \bar{\mathbf{f}}(k). \tag{4.3.5}$$

But from (4.1.20)

$$\delta\mathbf{x}(k) = \mathbf{u}_1(k)\delta\mathbf{x}(0) + \mathbf{v}_1(k)\delta\boldsymbol{\alpha}. \tag{4.3.6}$$

Substituting (4.3.6) in (4.3.5), we obtain

$$[\mathbf{D}_{\bar{\mathbf{x}}}(\mathbf{h})\mathbf{u}_1(k)]\,\delta\mathbf{x}(0) + [\mathbf{D}_{\bar{x}}(\mathbf{h})\mathbf{v}_1(k)]\,\delta\boldsymbol{\alpha} = \bar{\mathbf{f}}(k) \tag{4.3.7}$$

To clarify the structure of this equation, define

$$\begin{aligned}
\mathbf{H}_1 &= \mathbf{D}_{\bar{\mathbf{x}}}(\mathbf{h})\mathbf{u}_1(k) \in \mathbb{R}^{m\times n} \\
\mathbf{H}_2 &= \mathbf{D}_{\bar{\mathbf{x}}}(\mathbf{h})\mathbf{v}_1(k) \in \mathbb{R}^{m\times p} \\
\mathbf{H} &= [\mathbf{H}_1, \mathbf{H}_2] \in \mathbb{R}^{m\times(n+p)} \\
\boldsymbol{\zeta} &= (\boldsymbol{\zeta}_1^T, \boldsymbol{\zeta}_2^T)^T \in \mathbb{R}^{n+p}, \\
\boldsymbol{\zeta}_1 &= \delta\mathbf{x}(0) \in \mathbb{R}^n, \ \boldsymbol{\zeta}_2 = \delta\boldsymbol{\alpha} \in \mathbb{R}^p. \tag{4.3.8}
\end{aligned}$$

Then, using (4.3.8), Eq. (4.3.7) reduces to a linear system

$$\mathbf{H}\boldsymbol{\zeta} = \bar{\mathbf{f}}(k),$$ (4.3.9)

where there are m equations for the m-components of $z(k)$ in $\boldsymbol{\zeta}$, a vector with $(n+p)$ components. Since $m \neq (n+p)$, this equation can be solved only by the method of linear least squares. Refer to Lewis et al. (2006, Chap. 5).

If $m < (n+p)$, (4.3.9) constitutes an under-determined system wherein there are infinitely many solutions. An unique solution is obtained by invoking the method of Lagrangian multipliers (refer to Chap. 2). It can be verified that the solution in this case is given by

$$\boldsymbol{\zeta}_{LS} = \mathbf{H}^T \left(\mathbf{H}\mathbf{H}^T\right)^{-1} \mathbf{z}$$ (4.3.10)

$\boldsymbol{\zeta}_{LS}$ is computed by first solving $(\mathbf{H}\mathbf{H}^T)\mathbf{y} = \mathbf{z}$ and $\boldsymbol{\zeta}_{LS} = \mathbf{H}^T\mathbf{y}$. The Grammian $\mathbf{H}\mathbf{H}^T$ is clearly symmetric and it also positive definite when $\mathbf{H}$ is of full rank, that is $\text{Rank}(\mathbf{H}) = \min\{m, n+p\} = m$, in this case.

If $m > (n+p)$, then (4.3.9) is an over-determined system. In this case the system is often inconsistent and the solution $\boldsymbol{\zeta}_{LS}$ is obtained by solving

$$(\mathbf{H}^T\mathbf{H})\boldsymbol{\zeta} = \mathbf{H}^T\mathbf{z}, \ \text{ or } \ \boldsymbol{\zeta}_{LS} = (\mathbf{H}^T\mathbf{H})^{-1}\mathbf{H}^T\mathbf{z}.$$ (4.3.11)

In this case, the Grammian $(\mathbf{H}^T\mathbf{H})$ is symmetric and positive definite if the $\mathbf{H}$ is of full rank, that is $\text{Rank} = \min\{m, n+p\} = n+p$. There are at least three direct matrix decomposition algorithms for solving linear systems of the type $\mathbf{Ax} = \mathbf{b}$ with symmetric and positive definite matrix $\mathbf{A}$. These include (a) Cholesky factorization, (b) QR algorithm based on Gramm-Schmidt orthogonalization method and (c) singular value decomposition (SVD) based method. For details, refer to Lewis et al. (2006, Chap. 9). Of course, (4.3.10) and (4.3.11) can also be solved by using one of the several iterative methods including Jacobi, Gauss-Seidel, SQR, Conjugate-Gradient, to name a few. See Golub and Van Loan (2012).

4.3.2 Case 2: Multiple Observations

Let $\mathbf{z}(k_i)$ for $1 \leq i \leq N$ be the set of N observations available for data assimilation, where it is assumed that the observation times are ordered as

$$0 < k_1 < k_2 < \ldots < k_N$$ (4.3.12)

Analysis of the case is quite straightforward in that we need to apply the analysis in case 1 to each time k_i when the observation is available and simply collate all the equations into a single equation for the determination of the correction ζ to the control. Given this, we only indicate the key points in the analysis.

First, the forecast error relation in (4.3.1) and (4.3.2) takes the form

$$\bar{\mathbf{f}}(k_i) = \mathbf{z}(k_i) - \mathbf{h}\left(\bar{\mathbf{x}}(k_i)\right))) \tag{4.3.13}$$

and

$$\mathbf{f}(k_i) = \mathbf{z}(k_i) - \mathbf{h}\left(\mathbf{x}(k_i)\right))) \tag{4.3.14}$$

one for each time $k_i, 1 \leq i \leq N$. Recall that the induced perturbation $\delta\mathbf{x}(k)$ is still given by (4.1.20). Hence, the condition for the simultaneous annihilation of the forecast error in (4.3.13), in analogy with (4.3.5) gives rise to the following N conditions:

$$\mathbf{D}_{\bar{\mathbf{x}}(k_i)}(\mathbf{h})\delta\mathbf{x}(k_i) = \bar{\mathbf{f}}(k_i). \tag{4.3.15}$$

Substituting (4.3.6) with $k = k_i$ into (4.3.15) and simplifying we get

$$\mathbf{H}(i)\boldsymbol{\zeta} = \bar{\mathbf{f}}(k_i), 1 \leq i \leq N \tag{4.3.16}$$

where recall $\mathbf{H}(i)$ and $\boldsymbol{\zeta}$ are as defined in (4.3.8) with $k = k_i$. Now create a new matrix $\mathbf{H} \in \mathbb{R}^{Nm \times (n+p)}$ and $\bar{\mathbf{f}} \in \mathbb{R}^{Nm}$ where

$$\mathbf{H} = \begin{bmatrix} \mathbf{H}(1) \\ \mathbf{H}(2) \\ \vdots \\ \mathbf{H}(N) \end{bmatrix} \quad \text{and} \quad \bar{\mathbf{f}} = \begin{bmatrix} \bar{\mathbf{f}}(k_1) \\ \bar{\mathbf{f}}(k_2) \\ \vdots \\ \bar{\mathbf{f}}(k_N) \end{bmatrix}. \tag{4.3.17}$$

Consequently, the set of N equations in (4.3.16) can be reduced to one equation as

$$\mathbf{H}\boldsymbol{\zeta} = \bar{\mathbf{f}} \tag{4.3.18}$$

which is a standard linear least squares problem. If $Nm > (n+p)$, then it corresponds to an over determined case and the solution is given by

$$\boldsymbol{\zeta}_{LS} = (\mathbf{H}^T\mathbf{H})^{-1}\mathbf{H}^T\bar{\mathbf{f}} \tag{4.3.19}$$

which can be solved by one of several methods mentioned at the end of Sect. 4.3.1.

4.4 Exercises

Problem 4.1. By exploiting the structural similarity between (4.1.20) and the recurrence in Problem 2.3 in Chap. 2, verify the correctness of the expression in (4.1.21) and (4.1.22).

4.4.1 Demonstrations

Demonstration: 2-Mode Nonlinear Problem

In meteorology, the nonlinear advection equation

$$\frac{\partial u}{\partial t} + u \frac{\partial u}{\partial x} = 0, u(x, 0) = f(x), \qquad (AD.1)$$

is an equation introduced into fluid dynamics by J.M. Burgers, and accordingly sometimes referenced to as Burgers' equation (Platzman 1964). It has served dynamical meteorologists well. It has strong mathematical similarity to the barotropic vorticity equation, an equation that has been used in meteorology to follow long waves (waves of length significantly greater that the depth of the troposphere ($\sim$10 km)). This similarity has led to a host of studies that examine the interaction of waves. We will make use of the two-component spectral form of the equation to outline the FSM data assimilation problem.

This wave can be represented as the superposition of two waves, wave 1 and 2 as depicted in Fig. AD.1. Here we plot the wave amplitudes around a mid-latitudes circle in the Northern Hemisphere. Now the nonlinear advection equation applied to two such waves allows for interaction of the waves, one wave grows or decays at the expense of the other and vice versa while the energy (sum of squared amplitudes) remains fixed. The problem we present

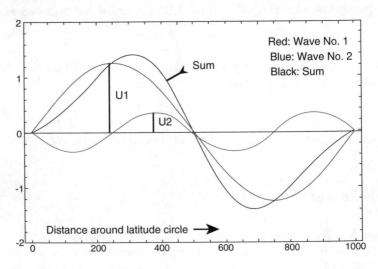

Fig. AD.1 Superposition of two sine waves

consists of two such waves (spectral form) that are governed by the truncated spectral form of solution (as found in Platzman 1964). We then treat the problem as a data assimilation problem (with solution by FSM) where observations of the wave amplitude (the sum of the two wave amplitudes) are made at fixed locations and a particular time around the latitude circle. The guess forecast (based on guess control) is then altered in accord with the FSM and this adjusted control then serves as a starting point for the subsequent forecast. The analytic form of governing equations for the truncated two-wave system follow:

$$\frac{du_1}{dt} = -\frac{1}{2}u_1 u_2,$$
$$\frac{du_2}{dt} = +\frac{1}{2}u_1^2,$$
(AD.2)

where $u(x, t) = -u_1(t)\sin(x) - u_2(t)\sin(2x)$, $0 \le x \le 2\pi$ and $u_1(0)$ and $u_2(0)$ are the given initial conditions. The distance around the latitude circle is scaled such that the nondimensional distance is 2π. These equations can be solved numerically by the Euler scheme.

$$u_1(k + 1) = -\frac{1}{2}u_1(k)u_2(k)\Delta t + u_1(k),$$
$$u_2(k + 1) = \frac{1}{2}[u_1(k)]^2 \Delta t + u_2(k),$$
(AD.3)

where $t_k = k\Delta t, k = 0, 1, 2 \ldots$

Over the spatial interval $0 \le x \le 2\pi$, we assume there are observations of the combined amplitude $(u_1 + u_2)$ at the following locations: $x = \pi/4$, $2\pi/4, 3\pi/4, 5\pi/4, 6\pi/4$ and $7\pi/4$.

In this case, the "observation matrix" is given by

$$\mathbf{H} = [h_{ij}] = \begin{bmatrix} -\sin\frac{\pi}{4} & -\sin\frac{2\pi}{4} \\ -\sin\frac{2\pi}{4} & -\sin\frac{4\pi}{4} \\ \vdots & \vdots \\ -\sin\frac{7\pi}{4} & -\sin\frac{14\pi}{4} \end{bmatrix}, \mathbf{H} \in \mathbb{R}^{6\times 2}$$
(AD.4)

and if the forecast at the observation time $t = T$ is

$$\mathbf{u}(T) = \begin{bmatrix} u_1(T) \\ u_2(T) \end{bmatrix}, \mathbf{u} \in \mathbb{R}^2,$$
(AD.5)

then the model counterpart is $HU \in \mathbb{R}^6$. The sum of the squared differences between the observation

$$\mathbf{z} = (z_1, z_2, \ldots, z_6)^T$$
(AD.6)

and the model counterpart drives the assimilation process. But for the FSM, we modify the guess forecast by including the sensitivity terms. In this case, we have two forecast variables and two elements of control. There are four elements in the sensitivity vector: $\frac{\partial u_1(t)}{\partial u_1(0)}$, $\frac{\partial u_1(t)}{\partial u_2(0)}$, $\frac{\partial u_2(t)}{\partial u_1(0)}$ and $\frac{\partial u_2(t)}{\partial u_2(0)}$. The forecast we use is the sum of the "guess" forecast, $\mathbf{u}(t)$, plus the adjustments due to sensitivity terms. In our case, the difference between forecast and observations uses the "augmented" forecast (guess forecast plus adjustment terms). This augmented forecast is

$$\hat{\mathbf{u}}(T) = \begin{bmatrix} u_1(T) + \frac{\partial u_1(t)}{\partial u_1(0)} \Delta u_1(0) + \frac{\partial u_1(t)}{\partial u_2(0)} \Delta u_2(0) \\ u_2(T) + \frac{\partial u_2(t)}{\partial u_1(0)} \Delta u_1(0) + \frac{\partial u_2(t)}{\partial u_2(0)} \Delta u_2(0) \end{bmatrix}. \tag{AD.7}$$

Thus, the problem reduces to finding adjustments to control such that the sum of the differences between the observations and the "augmented" model counterpart is minimized.

4.5 Notes and References

This chapter follows Lakshmivarahan and Lewis (2010) rather closely.

Chapter 5
Forecast Error Correction Using Optimal Tracking

The basic principle of forecast error correction using dynamic data assimilation (DDA) is to alter or move the model trajectory towards a given set of (noisy) observations in such a way that the weighted sum of squared forecast errors (which is also the square of the energy norm of the forecast errors) is a minimum. We have classified the various sources of forecast error in Sect. 1.5.2. A little reflection reveals that there are essentially two ways of altering the solution of a dynamical system: (1) changing the control consisting of the initial/boundary conditions and the parameters of the dynamical model, and (2) by adding an explicit external forcing (a form of state dependent control) which will in turn force the model solution towards the desired goal. The 4D-VAR and FSM based deterministic framework are designed to alter the model trajectory to correct the forecast errors by iteratively adjusting the control (initial/boundary conditions and parameters) using one of the well established algorithms for minimizing the square of the energy norm of the forecast errors. Refer to Chaps. 2 and 4 of this book and Lewis et al. (2006) for details. Identification of errors in the dynamics and associated adjustments go beyond correcting control. Generally, the correction terms added to the constraints force the model forecast to more closely fit the observations either empirically (so-called nudging process) or optimally (a process that involves a least squares fit of model to observation). Lakshmivarahan and Lewis (2013) have comprehensively viewed the work on nudging. In this chapter we address the process of optimally adjusting the constraints to fit the observations. The most powerful method to accomplish this task is Pontryagin's minimum principle (PMP). It is the centerpiece of this chapter. At the end of the chapter, we also consider an optimal method used in meteorology developed by Derber (1989) and applied to hurricane tracking by DeMaria and Jones (1993).

© Springer International Publishing Switzerland 2017
S. Lakshmivarahan et al., *Forecast Error Correction using Dynamic Data Assimilation*, Springer Atmospheric Sciences, DOI 10.1007/978-3-319-39997-3_5

Given a desired trajectory of the model which in the context of DDA is specified by the sequence of N noisy observations, the goal of the optimal tracking is to compute an external forcing in the form of a state dependent feedback control that in turn drives the model towards the desired trajectory which in turn reduces the forecast errors.

In Sect. 5.1 we provide a robust derivation of the solution of optimal tracking problem using PMP for a general nonlinear model. The intrinsic connection between PMP and 4D-VAR is explored in Sect. 5.2. The special structure of the optimal tracking solution for the linear case is derived in Sect. 5.3. An algorithm for quantifying model errors using PMP is derived in Sect. 5.4. Detailed illustration of the power of PMP based approach using a linear advection problem is given in Sect. 5.5. Notes and references are contained in Sect. 5.7. Appendix "Solution of the LOM in (5.5.8)" contains supplemental material that are key to the developments in the main body of the chapter.

5.1 Pontryagin's Minimum Principle (PMP) in Discrete Time

Let

$$\overline{\mathbf{x}}_{k+1} = \overline{\mathbf{M}}\left(\overline{\mathbf{x}}_k\right) \tag{5.1.1}$$

be the given discrete time nonlinear model dynamics where the map $\overline{\mathbf{M}} : \mathbb{R}^n \to \mathbb{R}^n$ with $\overline{\mathbf{M}}(\mathbf{x}) = \left(\overline{\mathbf{M}}_1(x), \overline{\mathbf{M}}_2(x), \ldots, \overline{\mathbf{M}}_n(x)\right)^T \in \mathbb{R}^n$ represents the imperfect model with $\overline{\mathbf{x}}_0$ as the initial conditions. For simplicity in notation, it is assumed that $\overline{\mathbf{M}}$ includes all the parameters and boundary conditions which are all intrinsic parts of the model dynamics.

Pontryagin's method calls for adding an external forcing term to the given model in (5.1.1) resulting in a nonlinear forced dynamics given by

$$\mathbf{x}_{k+1} = \mathbf{M}\left(\mathbf{x}_k, \mathbf{u}_k\right) \tag{5.1.2}$$

where $\mathbf{u}_k \in \mathbb{R}^p$ is the external forcing vector, $\mathbf{M} : \mathbb{R}^n \times \mathbb{R}^p$ with $\mathbf{M}(\mathbf{x}, \mathbf{u}) = (\mathbf{M}_1(\mathbf{x}, \mathbf{u}), \mathbf{M}_2(\mathbf{x}, \mathbf{u}), \ldots, \mathbf{M}_n(\mathbf{x}, \mathbf{u}))^T \in \mathbb{R}^n$ and $\mathbf{x}_0$ as the initial condition. In general $\mathbf{M}(\mathbf{x}, \mathbf{u})$ could be a nonlinear map in both $\mathbf{x}$ and $\mathbf{u}$, but for simplicity in implementation it is further assumed that $\mathbf{M}(\mathbf{x}, \mathbf{u})$ is linear and additive in $\mathbf{u}$ given by

$$\mathbf{M}(\mathbf{x}, \mathbf{u}) = \overline{\mathbf{M}}(\mathbf{x}) + \mathbf{B}\mathbf{u}, \tag{5.1.3}$$

and

$$\mathbf{x}_{k+1} = \overline{\mathbf{M}}(\mathbf{x}_k) + \mathbf{B}u_k, \tag{5.1.4}$$

with $\mathbf{B} \in \mathbb{R}^{n \times p}$ with $1 \le p \le n$. When $p = 1$ and $\mathbf{B} = (1, 1, \ldots, 1)^T \in \mathbb{R}^n$, $\mathbf{u}_k \in \mathbb{R}$ is a scalar and the same force $\mathbf{u}_k$ is applied to every component $\overline{M}_i(\mathbf{x}_k)$ for $1 \le i \le n$. At the other extreme, when $p = n$ and $\mathbf{B} = \mathbb{I} \in \mathbb{R}^{n \times n}$ the identity matrix, then the ith component $\mathbf{u}_{k,i}$ of the forcing vector $\mathbf{u}_k$ is applied to the ith component of the model map $\overline{M}_i(\mathbf{x}_k)$. Thus, the larger the value of p, better the chance to steer the individual components of the model state towards the desired regions of the state space.

Let observations be given by

$$\mathbf{z}_k = \mathbf{h}(\mathbf{x}_k) + \mathbf{v}_k \tag{5.1.5}$$

where $\mathbf{h} : \mathbb{R}^n \to \mathbb{R}^m$ is called the forward operator that maps the model state $\mathbb{R}^n$ into the observation space $\mathbb{R}^m$, $\mathbf{z}_k \in \mathbb{R}^m$ is the observation vector and $\mathbf{v}_k \sim N(0, \mathbf{R}_k)$ is the observation (Gaussian) noise vector with mean zero and known covariance $\mathbf{R}_k$.

Let $\langle \mathbf{a}, \mathbf{b} \rangle$ denote the inner product $\mathbb{R}^n$. Let $\mathbf{x}(0 : k)$ denote the sequence of state vectors $\{\mathbf{x}_0, \mathbf{x}_1, \ldots, \mathbf{x}_k\}$ and likewise $\mathbf{u}(0 : k)$ denote the sequence of control vectors. Define a performance measure

$$J(\mathbf{u}(0 : N - 1)) = \sum_{k=0}^{N-1} V_k, \tag{5.1.6}$$

where

$$V_k = V_k(\mathbf{x}_k, \mathbf{z}_k, \mathbf{u}_k) = V_k^0(\mathbf{x}_k, \mathbf{z}_k) + V_k^c(\mathbf{u}_k), \tag{5.1.7}$$

with

$$V_k^0(\mathbf{x}_k, \mathbf{z}_k) = \frac{1}{2} \langle \mathbf{e}_k, \mathbf{R}^{-1} \mathbf{e}_k \rangle, \tag{5.1.8}$$

$$V_k^c(\mathbf{u}_k) = \frac{1}{2} \langle \mathbf{e}_k, \mathbf{C} \mathbf{u}_k \rangle, \tag{5.1.9}$$

and

$$\mathbf{e}_k = \mathbf{z}_k - \mathbf{h}(\mathbf{x}_k) \tag{5.1.10}$$

is the forecast error.

Clearly, V_k^0 is the square of the energy norm of the forecast error and V_k^c is the square of the energy norm of the control vector $\mathbf{u}_k$ where $\mathbf{C} \in \mathbb{R}^{p \times p}$ is a given symmetric and positive definite matrix.

Our goal is to find the optimal control sequence $\mathbf{u}(0 : N - 1)$ that minimizes J in (5.1.6) when the model states sequence $\mathbf{x}(0 : N)$ are constrained by the dynamical model equation in (5.1.4). This strong constraint minimization problem is reformulated as an unconstrained minimization problem by invoking the Lagrangian multiplier technique. To this end, first define a Lagrangian

$$L = L(\mathbf{u}(0:N-1), \boldsymbol{\lambda}(1:N))$$

$$= \sum_{k=0}^{N-1} \{V_k + \langle \boldsymbol{\lambda}_{k+1}, \mathbf{M}(\mathbf{x}_k, \mathbf{u}_k) - \mathbf{x}_{k+1}\rangle\}, \tag{5.1.11}$$

where $\mathbf{M}(\mathbf{x}, \mathbf{u})$ is given by (5.1.3), $\boldsymbol{\lambda}_k \in \mathbb{R}^n$ is the kth (undetermined) Lagrangian multiplier vector and $\boldsymbol{\lambda}(1:N)$ denotes the sequence $\{\boldsymbol{\lambda}_1, \boldsymbol{\lambda}_2, \ldots, \boldsymbol{\lambda}_N\}$ of Lagrangian multiplier vectors. Instead of minimizing L in (5.1.11), Pontryagin's approach calls for first rewriting L in terms of the Hamiltonian function defined by

$$H_k = H_k(\mathbf{x}_k, \mathbf{u}_k, \boldsymbol{\lambda}_k) \quad = V_k + \langle \boldsymbol{\lambda}_{k+1}, \mathbf{M}(\mathbf{x}_k, \mathbf{u}_k)\rangle. \tag{5.1.12}$$

Now substituting (5.1.12) in (5.1.11) and simplifying we get

$$L = \sum_{k=0}^{N-1} [H_k - \langle \boldsymbol{\lambda}_{k+1}, \mathbf{x}_{k+1}\rangle]. \tag{5.1.13}$$

It is convenient to rewrite L as

$$L = H_0 - \langle \boldsymbol{\lambda}_N, \mathbf{x}_N\rangle + \sum_{k=0}^{N} [H_k - \langle \boldsymbol{\lambda}_k, \mathbf{x}_k\rangle]. \tag{5.1.14}$$

As a first step towards minimizing L, let us compute the first variation δL in L induced by increments $\delta \mathbf{x}_k$ for $0 \leq k \leq N$, $\delta \mathbf{u}_k$ in $\mathbf{u}_k$ for $0 \leq k \leq N-1$ and increments $\delta \boldsymbol{\lambda}_k$ for $1 \leq k \leq N$.

Since H_k is a scalar function of the vectors $\mathbf{x}_k, \mathbf{u}_k$ and $\boldsymbol{\lambda}_{k+1}$, from the first principles (refer to Appendix) it follows that

$$\delta L = \langle \nabla_{x_0} H_0, \delta \mathbf{x}_0\rangle + \langle \nabla_{u_0} H_0, \delta \mathbf{u}_0\rangle + \langle \nabla_{\lambda_1} H_0, \delta \boldsymbol{\lambda}_1\rangle$$

$$- \langle \delta \boldsymbol{\lambda}_N, \mathbf{x}_N\rangle - \langle \boldsymbol{\lambda}_N, \delta \mathbf{x}_N\rangle$$

$$+ \sum_{k=1}^{N-1} \langle \nabla_{x_k} H_k, \delta \mathbf{x}_k\rangle - \sum_{k=1}^{N-1} \langle \boldsymbol{\lambda}_k, \delta \mathbf{x}_k\rangle$$

$$+ \sum_{k=1}^{N-1} \langle \nabla_{u_k} H_k, \delta \mathbf{u}_k\rangle$$

$$+ \sum_{k=1}^{N-1} \langle \nabla_{\lambda_{k+1}} H_k, \delta \boldsymbol{\lambda}_{k+1}\rangle - \sum_{k=1}^{N-1} \langle \delta \boldsymbol{\lambda}_k, \mathbf{x}_k\rangle.$$

By collecting like terms and simplifying δL becomes

$$\delta L = \langle \nabla_{\mathbf{x}_0} H_0, \delta \mathbf{x}_0 \rangle - \langle \lambda_N, \delta \mathbf{x}_N \rangle$$

$$+ \sum_{k=0}^{N-1} \langle \nabla_{\mathbf{u}_k} H_k, \delta \mathbf{u}_k \rangle$$

$$- \sum_{k=1}^{N} \langle (\nabla_{\lambda_k} H_{k-1} - \mathbf{x}_k), \delta \lambda_k \rangle$$

$$+ \sum_{k=1}^{N-1} \langle (\nabla_{\mathbf{x}_k} H_k - \lambda_k), \delta \mathbf{x}_k \rangle. \tag{5.1.15}$$

where $\nabla_{\mathbf{x}_k} H_k \in \mathbb{R}^n$, $\nabla_{\mathbf{u}_k} H_k \in \mathbb{R}^p$ and $\nabla_{\lambda_k} H_{k-1} \in \mathbb{R}^n$. Recall that δL must be zero at the minimum and in view of the arbitrariness of $\delta \mathbf{x}_k$, $\delta \mathbf{u}_k$ and $\delta \lambda_k$, we obtain the following set of necessary conditions for the minimum of L.

5.1.1 Condition 1: Model Dynamics

The fourth term on the right hand side of (5.1.15) is zero when

$$\mathbf{x}_k = \nabla_{\lambda_k} H_{k-1} = \frac{\partial H_{k-1}}{\partial \lambda_k} \quad \text{for } 1 \leq k \leq N. \tag{5.1.16}$$

Now computing the gradient of H_{k-1} in (5.1.12) with respect to λ_k and using (5.1.3) in (5.1.16), we obtain the first set of necessary conditions, namely

$$\mathbf{x}_k = \mathbf{M}(\mathbf{x}_{k-1}, \mathbf{u}_{k-1}) = \overline{\mathbf{M}}(\mathbf{x}_{k-1}) + \mathbf{B}\mathbf{u}_{k-1}. \tag{5.1.17}$$

Stated in other words, Pontryagin's method dictates that the sequence of states $\mathbf{x}_0, \mathbf{x}_1, \ldots, \mathbf{x}_N$ must arise as a solution of the model dynamics (5.1.4). That is, the model represents a strong constraint.

5.1.2 Condition 2: Co-State or Adjoint Dynamics

The fifth term on the right hand side of (5.1.15) vanishes when

$$\lambda_k = \nabla_{\mathbf{x}_k} H_k = \frac{\partial H_k}{\partial \mathbf{x}_k} \quad \text{for } 1 \leq k \leq N - 1. \tag{5.1.18}$$

Now computing the gradient of H_k in (5.1.12) and substituting in (5.1.18), we obtain the second set of necessary conditions given by

$$\boldsymbol{\lambda}_k = \mathbf{D}_{\mathbf{x}_k}^T(\overline{\mathbf{M}})\boldsymbol{\lambda}_{k+1} + \nabla_{\mathbf{x}_k} V_k, \tag{5.1.19}$$

where the model Jacobian

$$\mathbf{D}_{\mathbf{x}}(\overline{\mathbf{M}}) = \left[\frac{\partial \overline{\mathbf{M}}_i}{\partial x_j}\right] \in \mathbb{R}^{n \times m}. \tag{5.1.20}$$

From (5.1.7) and (5.1.8), the gradient of V_k with respect to $\mathbf{x}_k$ is given by

$$\nabla_{\mathbf{x}_k} V_k = \nabla_{\mathbf{x}_k} V_k^0 = -\mathbf{D}_{\mathbf{x}_k}^T(\mathbf{h})\mathbf{R}^{-1}\left[\mathbf{z}_k - \mathbf{h}(\mathbf{x}_k)\right] = -\mathbf{D}_{\mathbf{x}_k}^T(\mathbf{h})\mathbf{R}^{-1}\mathbf{e}_k = \mathbf{f}_k \in \mathbb{R}^n, \tag{5.1.21}$$

which is the nonlinear forecast error when viewed from the model space where

$$\mathbf{D}_{\mathbf{x}}(\mathbf{h}) = \left[\frac{\partial \overline{\mathbf{h}}_i}{\partial x_j}\right] \in \mathbb{R}^{m \times n} \tag{5.1.22}$$

is the Jacobian of the forward operator $\mathbf{h}(\mathbf{x})$.

Combining (5.1.19) and (5.1.22), we obtain the well known adjoint or the co-state dynamics (Lewis et al. 2006).

$$\boldsymbol{\lambda}_k = \mathbf{D}_{x_k}^T(\overline{\mathbf{M}})\boldsymbol{\lambda}_{k+1} + \mathbf{f}_k \tag{5.1.23}$$

is the second set of necessary conditions.

5.1.3   Condition 3: Stationarity Condition

From the third term on the right hand side of (5.1.15), we readily obtain the third necessary condition as

$$0 = \nabla_{\mathbf{u}_k} H_k = \frac{\partial H_k}{\partial \mathbf{u}_k} \quad \text{for } 0 \le k \le N - 1. \tag{5.1.24}$$

Again computing the gradient of H_k in (5.1.12), we obtain

$$0 = \nabla_{\mathbf{u}_k} V_k + \mathbf{D}_{\mathbf{u}_k}^T(\mathbf{M})\boldsymbol{\lambda}_{k+1} \tag{5.1.25}$$

and from (5.1.7)–(5.1.19) it follows that

$$\nabla_{\mathbf{u}_k} V_k = \nabla_{\mathbf{u}_k} V_k^C = \mathbf{C}\mathbf{u}_k. \tag{5.1.26}$$

From (5.1.3) we get

$$\mathbf{D_u(M)} = \mathbf{B}. \tag{5.1.27}$$

And by combining (5.1.25)–(5.1.27), we obtain the structure of the optimal control for $0 \le k \le N - 1$ as

$$\mathbf{u}_k = -\mathbf{C}^{-1}\mathbf{B}^T\boldsymbol{\lambda}_{k+1}, \tag{5.1.28}$$

which is well defined since the matrix $\mathbf{C}$ is assumed to be symmetric and positive definite.

5.1.4 Condition 4: Boundary Conditions

We are left with the first two terms on the right hand side of (5.1.15) and the conditions that force these two terms to zero provide us with the required boundary conditions for computing the optimal control sequence $\mathbf{u}(0 : N - 1)$.

Recall that $\mathbf{x}_0$ is given but $\mathbf{x}_N$ is assumed to be free, Hence $\delta\mathbf{x}_0 = 0$ and $\delta\mathbf{x}_N$ are arbitrary. Thus, the first term on the right hand side of (5.1.15) is automatically zero. The second term is forced to zero by setting

$$\boldsymbol{\lambda}_N = 0. \tag{5.1.29}$$

The above developments naturally lead to the following:

5.1.4.1 A Framework for Optimal Tracking

The structure of the optimal control sequence $\mathbf{u}_k$ obtained by solving the stationarity condition (5.1.24) is given in (5.1.28). Substituting the expression for $\mathbf{u}_k$ in the model dynamics (5.1.4) delivers the evolution of the optimal trajectory that is given by

$$\mathbf{x}_{k+1} = \overline{\mathbf{M}}(\mathbf{x}_k) - \mathbf{BC}^{-1}\mathbf{B}^T\boldsymbol{\lambda}_{k+1}, \tag{5.1.30}$$

where $\mathbf{x}_0$ is the initial condition. The backward evolution of the Lagrangian multipliers is given by the co-state or the adjoint dynamics in (5.1.23) repeated here for convenience:

$$\boldsymbol{\lambda}_k = \mathbf{D}_{\mathbf{x}_k}^T(\overline{\mathbf{M}})\boldsymbol{\lambda}_{k+1} + \mathbf{f}_k, \tag{5.1.31}$$

with the final condition $\lambda_N = 0$. Clearly by solving the nonlinear coupled two point boundary value problem (TPBVP) (5.1.30) and (5.1.31), we obtain the optimal trajectory, where the model dynamics (5.1.30) evolves forward in time from the given initial condition $\mathbf{x}_0$ and the co-state/adjoint dynamics is solved backward in time from $\lambda_N = 0$.

A number of observations are in order:

1. In the standard Lagrangian framework, the Euler-Lagrange necessary condition for the minimum of L is given by a nonlinear second order equation. The importance of the Hamiltonian formulation stems from the simplicity and conciseness of necessary condition using the two first order equations arising from (5.1.16) and (5.1.23), involving the state and the adjoint dynamics.
2. The TPBVP in (5.1.30) and (5.1.31) admits a closed form solution only when the given model in (5.1.1) and the forward operator $\mathbf{h}(\cdot)$ in (5.1.5) are both linear. In all of the other cases (5.1.30) and (5.1.31) have to be solved using clever numerical methods. A number of numerical methods for solving (5.1.30)–(5.1.31) have been developed in the literature, a sampling of the methods can be found in Roberts and Shipman (1972), Keller (1976), Polak (2012), and Bryson and Ho (1999). The closed form solution for the linear case is derived in Sect. 5.3.

5.2 Connection to 4D-VAR

In this section we provide a derivation of the 4D-VAR method as a special case of the developments in Sect. 5.1. To this end, consider the unforced version of the model equations given by (5.1.4)

$$\mathbf{x}_{k+1} = \overline{\mathbf{M}}(\mathbf{x}_k), \tag{5.2.1}$$

with $\mathbf{x}_0$ as the initial condition. Since there is no external forcing in (5.2.1), the only way to move the model trajectory to minimize the energy norm of the forecast error in (5.1.10) is by changing the initial condition $\mathbf{x}_0$.

Consider the special case of the cost function in (5.1.6) obtained by setting $\mathbf{u}_k \equiv 0$ that leads to a new cost function

$$\overline{J}(\mathbf{x}_0) = \sum_{k=0}^{N-1} \overline{V}_k, \tag{5.2.2}$$

where

$$\overline{V}_k = \overline{V}(\mathbf{x}_k, \mathbf{z}_k) = V_k^0(\mathbf{x}_k, \mathbf{z}_k), \tag{5.2.3}$$

given in (5.1.8) and (5.1.10). Our goal is to minimize $\bar{J}(x_0)$ subject to the model equation (5.2.1) as the strong constraint. To this end, derive a new Lagrangian

$$\bar{L} = \bar{L}\left(\mathbf{x}_0,\ \bar{\lambda}(1:N)\right) = \bar{J}(x_0) + \sum_{k=0}^{N-1} \langle \bar{\lambda}_{k+1}, \overline{\mathbf{M}}(\mathbf{x}_k) - \mathbf{x}_{k+1} \rangle$$

$$= \sum_{k=0}^{N-1} [\overline{V}_{k+1} + \langle =_{k+1}, \overline{\mathbf{M}}(\mathbf{x}_k) - \mathbf{x}_{k+1} \rangle] \tag{5.2.4}$$

where $\bar{\lambda}(1:N) = \left(\bar{\lambda}_1, \bar{\lambda}_2, \ldots, \bar{\lambda}_N\right)$ is a sequence of undetermined Lagrangian multipliers with $\lambda_k \in \mathbb{R}^n$ for $1 \le k \le N$.

As in Sect. 5.1, now define the associated Hamiltonian function

$$\overline{H}_k = \overline{V}_k + \langle \bar{\lambda}_{k+1}, \overline{\mathbf{M}}(\mathbf{x}_k) - \mathbf{x}_{k+1} \rangle. \tag{5.2.5}$$

Substituting (5.2.5) into (5.2.4) and simplifying, we obtain

$$\bar{L} = \overline{H}_0 + \langle \bar{\lambda}_N, \mathbf{x}_N \rangle + \sum_{k=1}^{N-1} \left[\overline{H}_k - \langle \bar{\lambda}_N, \mathbf{x}_N \rangle \right], \tag{5.2.6}$$

which is very similar to L in (5.1.14) except that $\bar{L}$ does not depend on the external control $\mathbf{u}_k, 0 \le k \le N-1$.

As a first step in minimizing $\bar{L}$, let $\delta\bar{L}$ be the induced first variation in $\bar{L}$ obtained from the variations $\delta\mathbf{x}_k$ in $\mathbf{x}_k, 0 \le k \le N$ and $\delta\bar{\lambda}_k$ for $1 \le k \le N$. Clearly, $\delta\bar{L}$ is obtained from δL in (5.1.15) by setting the $\delta\mathbf{u}_k$ terms to zero. Thus

$$\bar{\delta}L = \langle \nabla_{\mathbf{x}_0}\overline{H}_0, \delta\mathbf{x}_0 \rangle - \langle \bar{\lambda}_N, \delta\mathbf{x}_N \rangle + \sum_{k=1}^{N-1} \langle \nabla_{\lambda_k}\overline{H}_{k-1} - \mathbf{x}_k, \delta\bar{\lambda}_k \rangle$$

$$+ \sum_{k=1}^{N-1} \langle \nabla_{\mathbf{x}_k}\overline{H}_k - \bar{\lambda}_k, \delta\mathbf{x}_k \rangle. \tag{5.2.7}$$

The set of necessary conditions for the minimum of $\bar{L}$ are given below.

5.2.1 Condition 1: Model Dynamics

Vanishing of the third term on the right hand side of (5.2.7) when $\delta\bar{\lambda}_k$ are arbitrary leads to

$$\mathbf{x}_k = \nabla_{\bar{\lambda}_k}\overline{H}_{k-1}, \tag{5.2.8}$$

which using (5.2.5) reduces to model equations

$$\mathbf{x}_k = \overline{\mathbf{M}}(\mathbf{x}_{k-1}). \tag{5.2.9}$$

5.2.2   Condition 2: Co-State or Adjoint Dynamics

Since $\delta\mathbf{x}_k$ are arbitrary, the fourth term on the right hand side of (5.2.7) vanishes when

$$\overline{\lambda}_k = \nabla_{\mathbf{x}_k}\overline{H}_k, \tag{5.2.10}$$

which using (5.2.5) defines the adjoint dynamics give by

$$\overline{\lambda}_k = \mathbf{D}_{\mathbf{x}_k}^T(\overline{\mathbf{M}})\overline{\lambda}_{k+1} + \mathbf{f}_k \tag{5.2.11}$$

where $\mathbf{f}_k$ is given by (5.1.21).

5.2.3   Condition 3: Boundary Conditions

Since $\mathbf{x}_N$ is free, $\delta\mathbf{x}_N$ is arbitrary. Hence the second term on the right hand side of (5.2.7) is zero when

$$\overline{\lambda}_N = 0. \tag{5.2.12}$$

Combining (5.2.8) and (5.2.10) and (5.2.11), it follows that

$$\delta\overline{L} = \langle \nabla_{x_0}\overline{H}_0, \delta\mathbf{x}_0 \rangle. \tag{5.2.13}$$

From first principles, it is immediate that

$$\frac{\partial\overline{L}}{\partial x_0} = \nabla_{x_0}\overline{H}_0 = \mathbf{D}_{\mathbf{x}_0}^T(\overline{\mathbf{M}})\lambda_1 + \mathbf{f}_0. \tag{5.2.14}$$

The above development naturally gives rise to the 4D-VAR algorithm (Lewis et al. 2006) given in Algorithm 5.1

Algorithm 5.1 4D-VAR algorithm (Lewis et al. 2006)

> **Step 1:** Pick an $\mathbf{x}_0$ and compute the model solution $\mathbf{x}_k$
> for $0 \leq k \leq N$ using (5.2.1)
> **Step 2:** Using the observations $\mathbf{z}_k$, compute
> $\mathbf{f}_k = \mathbf{D}_{\mathbf{x}_k}^T(\overline{\mathbf{M}})R^{-1}[\mathbf{h}(\mathbf{x}_k) - \mathbf{z}_k]$
> **Step 3:** Compute $\overline{\boldsymbol{\lambda}}_1$ by running adjoint dynamics
> (5.2.11) backward in time starting from $\overline{\boldsymbol{\lambda}}_N = 0$
> **Step 4:** Compute $\partial\overline{L}/\partial\mathbf{x}_0$ using (5.2.14). If $\partial\overline{L}/\partial\mathbf{x}_0$
> is close to zero, we are done. Else use $\partial\overline{L}/\partial\mathbf{x}_0$ is a minimization algorithm find a new
> initial condition $\mathbf{x}_0$ until $\partial\overline{L}/\partial\mathbf{x}_0$ is close to zero.

5.3   Optimal Tracking: Linear Case

In this section we apply the PMP framework developed in Sect. 5.1 to the special case when (a) the dynamics is linear, (b) the observations are linear functions of the state and (c) the cost function is quadratic in the state and control. It turns out that a key impact of this linearity assumption is that by exploiting the hidden intrinsic affine relation between the co-state or adjoint variable and the state variable, we can obtain the optimal control and the optimal trajectory by solving two simpler initial value problems as opposed to solving one difficult two point boundary value problem.

Let

$$\mathbf{x}_{k+1} = \mathbf{M}\mathbf{x}_k + \mathbf{B}\mathbf{u}_k \qquad (5.3.1)$$

be the forced linear model when $\mathbf{x}_k \in \mathbb{R}^n$, $\mathbf{M} \in \mathbb{R}^{n \times m}$ is the one step state transition matrix, $\mathbf{u}_k \in \mathbb{R}^p$ is the control vector and $\mathbf{B} \in \mathbb{R}^{n \times p}$. Since $\overline{\mathbf{M}}(\mathbf{x}) = \mathbf{M}\mathbf{x}$, it is immediate that $\mathbf{D}_{\mathbf{x}}(\overline{\mathbf{M}}) = \mathbf{M}$.

Let the observations be given by

$$\mathbf{z}_k = \mathbf{H}\mathbf{x}_k + \mathbf{v}_k, \qquad (5.3.2)$$

where $\mathbf{z}_k \in \mathbb{R}^m$, $\mathbf{H} \in \mathbb{R}^{m \times n}$ and $\mathbf{v}_k \sim N(0, R)$ is the observation noise vector which has a multivariate Gaussian distribution with mean zero and known positive definite covariance matrix $\mathbf{R} \in \mathbb{R}^{m \times m}$. Since $\mathbf{h}(\mathbf{x}) = \mathbf{H}\mathbf{x}$, it follows that $\mathbf{D}_{\mathbf{x}}(\mathbf{h}) = \mathbf{H}$.

We consider the same quadratic cost function defined in (5.1.8)–(5.1.10). By substituting

$$\overline{\mathbf{M}}(\mathbf{x}) = \mathbf{M}\mathbf{x} \quad \text{and} \quad \mathbf{M}(\mathbf{x}, \mathbf{u}) = \mathbf{M}\mathbf{x} + \mathbf{B}\mathbf{u} \qquad (5.3.3)$$

in the expression for the Lagrangian L in (5.1.11) it can easily be verified that the necessary conditions take the following form.

5.3.1   Structure of the Optimal Control

From the stationary conditions in (5.1.24)–(5.1.28), it follows that the optimal control in this linear case is also given by

$$\mathbf{u}_k = -\mathbf{C}^{-1}\mathbf{B}^T\boldsymbol{\lambda}_{k+1} \tag{5.3.4}$$

where C is defined in (5.1.9).

5.3.2   The TPBVP

Substituting (5.3.3) and (5.3.4) in (5.1.30)–(5.1.31), the optimal trajectory $\mathbf{x}_k$ is obtained by solving the linear TPBVP given by the pair of equations

$$\mathbf{x}_{k+1} = \mathbf{M}\mathbf{x}_k - \left(\mathbf{B}\mathbf{C}^{-1}\mathbf{B}^T\right)\boldsymbol{\lambda}_{k+1}, \tag{5.3.5}$$

$$\boldsymbol{\lambda}_k = \mathbf{M}^T\boldsymbol{\lambda}_{k+1} + \mathbf{f}_k \tag{5.3.6}$$

where, since $\mathbf{h}(\mathbf{x}) = \mathbf{H}\mathbf{x}$ and $\mathbf{D}_x(\mathbf{h}) = \mathbf{H}$,

$$\mathbf{f}_k = \mathbf{H}^T\mathbf{R}^{-1}[\mathbf{H}\mathbf{x}_k - \mathbf{z}_k], \tag{5.3.7}$$

with $\mathbf{x}_0$ as the initial condition for (5.3.5) and $\boldsymbol{\lambda}_N = 0$ as the final condition for (5.3.6). To simplify the notation, define

$$\mathbf{E} = \mathbf{B}\mathbf{C}^{-1}\mathbf{B}^T \in \mathbb{R}^{n\times m}, \ \mathbf{F} = \mathbf{H}^T\mathbf{R}^{-1}\mathbf{H} \in \mathbb{R}^{n\times n}, \ \mathbf{W} = \mathbf{H}^T\mathbf{R}^{-1} \in \mathbb{R}^{n\times m}. \tag{5.3.8}$$

Then, using (5.3.8), the above linear TPBVP takes the form

$$\begin{bmatrix} \mathbf{x}_{k+1} \\ \boldsymbol{\lambda}_k \end{bmatrix} = \begin{bmatrix} \mathbf{M} & -\mathbf{E} \\ \mathbf{F} & \mathbf{M}^T \end{bmatrix} \begin{bmatrix} \mathbf{x}_k \\ \boldsymbol{\lambda}_{k+1} \end{bmatrix} + \begin{bmatrix} \mathbf{0} \\ -\mathbf{W} \end{bmatrix} z_k. \tag{5.3.9}$$

While in general it is difficult to solve a TPBVP, it turns out that the above special linear TPBVP can be converted into two initial value problems which are far easier to solve. The trick lies in uncovering a hidden affine relation between $\boldsymbol{\lambda}_k$ and $\mathbf{x}_k$.

5.3.3   Affine Relation Between $\boldsymbol{\lambda}_k$ and $\mathbf{x}_k$

Since $\boldsymbol{\lambda}_N = 0$, setting $k = N - 1$ in (5.3.9), we obtain the first of a series of affine relations given by

$$\boldsymbol{\lambda}_{N-1} = \mathbf{F}\mathbf{x}_{N-1} - \mathbf{W}\mathbf{z}_{N-1}. \tag{5.3.10}$$

This is the basis for our inductive generalization. Now using (5.3.10) in (5.3.9) and after simplifying for $k = N - 1$, we get

$$\lambda_{N-2} = \mathbf{F}\mathbf{x}_{N-2} - \mathbf{M}^T\mathbf{x}_{N-1} - \mathbf{M}^T\mathbf{W}\mathbf{z}_{N-1} - \mathbf{W}\mathbf{z}_{N-2}. \qquad (5.3.11)$$

Again using (5.3.10) in (5.3.9), it follows that

$$\mathbf{x}_{N-1} = \mathbf{M}\mathbf{x}_{N-2} - \mathbf{E}\lambda_{N-1} = \mathbf{M}\mathbf{x}_{N-2} - \mathbf{E}\mathbf{F}\mathbf{x}_{N-1} + \mathbf{E}\mathbf{W}\mathbf{z}_{N-1},$$

which leads to

$$(\mathbf{I} + \mathbf{E}\mathbf{F})\mathbf{x}_{N-1} = \mathbf{M}\mathbf{x}_{N-2} + \mathbf{E}\mathbf{W}\mathbf{z}_{N-1}. \qquad (5.3.12)$$

Substituting (5.3.12) in (5.3.11), we obtain the required affine relation as

$$\lambda_{N-2} = [\mathbf{F} + \mathbf{M}^T(\mathbf{I} + \mathbf{E}\mathbf{F})^{-1}\mathbf{M}]\mathbf{x}_{N-2}]$$
$$+ \mathbf{M}^T[(\mathbf{I} + \mathbf{E}\mathbf{F})^{-1}\mathbf{E} - \mathbf{I}]\mathbf{W}\mathbf{z}_{N-1} - \mathbf{W}\mathbf{z}_{N-2}. \qquad (5.3.13)$$

Continuing this argument, it is evident that we can express

$$\lambda_k = \mathbf{P}_k\mathbf{x}_k - \mathbf{g}_k \qquad (5.3.14)$$

for some matrix $\mathbf{P}_k \in \mathbb{R}^{n\times n}$ and vector $\mathbf{g}_k \in \mathbb{R}^n$.

5.3.4 Conversion of TPBVP to Two Initial Value Problems

Substituting (5.3.14) in the state equation (5.3.9) and rearranging we get

$$\mathbf{x}_{k+1} = (\mathbf{I} + \mathbf{E}\mathbf{P}_{k+1})^{-1}(\mathbf{M}\mathbf{x}_k + \mathbf{E}\mathbf{g}_{k+1}). \qquad (5.3.15)$$

Again substituting (5.3.14) and (5.3.15) in the co-state dynamics in (5.3.9) and simplifying, it follows that

$$\left[-\mathbf{P}_k + \mathbf{M}^T\mathbf{P}_{k+1}(\mathbf{I} + \mathbf{E}\mathbf{P}_{k+1})^{-1}\mathbf{M} + \mathbf{F}\right]\mathbf{x}_k$$
$$+ \left[\mathbf{g}_k + \mathbf{M}^T\mathbf{P}_{k+1}(\mathbf{I} + \mathbf{E}\mathbf{P}_{k+1})^{-1}\mathbf{E}\mathbf{g}_{k+1} - \mathbf{M}^T\mathbf{g}_{k+1} - \mathbf{W}\mathbf{z}_k\right] = 0. $$
$$(5.3.16)$$

The only way this relation can hold for all k and all $\mathbf{x}_k$ is that both the coefficient of $\mathbf{x}_k$ and the term independent of $\mathbf{x}_k$ must vanish simultaneously for all k. This observation immediately leads to two new dynamics of backward evolution of $\mathbf{P}_k$ and $\mathbf{g}_k$ given by

$$\mathbf{P}_k = \mathbf{M}^T\mathbf{P}_{k+1}(\mathbf{I} + \mathbf{E}\mathbf{P}_{k+1})^{-1}\mathbf{M} + \mathbf{F}, \qquad (5.3.17)$$

which is a nonlinear matrix Riccati equation which is independent of the observations and

$$\mathbf{g}_k = \mathbf{M}^T \mathbf{g}_{k+1} - \mathbf{M}^T \mathbf{P}_{k+1} (\mathbf{I} + \mathbf{E} \mathbf{P}_{k+1})^{-1} \mathbf{E} \mathbf{g}_{k+1} + \mathbf{W} \mathbf{z}_k. \tag{5.3.18}$$

Since $\boldsymbol{\lambda}_N = 0$ and $\mathbf{x}_N$ is arbitrary, from (5.3.14) we obtain the two final conditions

$$\mathbf{P}_N = 0 \text{ and } \mathbf{g}_N = 0. \tag{5.3.19}$$

Thus, we can compute the sequence of matrix $\{\mathbf{P}_k\}$ and vector $\{\mathbf{g}_k\}$ by solving the two initial value problems in (5.3.17)–(5.3.18) using the final conditions (5.3.19). It is important to note that, since $\mathbf{P}_k$ are independent of the observations, $\{\mathbf{P}_k\}$ can be computed and stored.

5.3.5 Structure of the Off-Line Optimal Control

Substituting (5.3.14) in (5.3.4), we see that

$$\mathbf{u}_k = -\mathbf{C}^{-1} \mathbf{B}^T [\mathbf{P}_{k+1} \mathbf{x}_{k+1} - \mathbf{g}_{k+1}]. \tag{5.3.20}$$

Using the dynamics in (5.3.1) in (5.3.20) we obtain

$$\mathbf{u}_k = -\mathbf{C}^{-1} \mathbf{B}^T [\mathbf{P}_{k+1} (\mathbf{M} \mathbf{x}_k + \mathbf{B} u_k) - \mathbf{g}_{k+1}]. \tag{5.3.21}$$

Premultiplying both sides by $\mathbf{C}$ and rearranging we get

$$\mathbf{u}_k = -\mathbf{K}_k \mathbf{x}_k + \mathbf{G}_k \mathbf{g}_{k+1}, \tag{5.3.22}$$

where the state feedback gain $\mathbf{K}_k \in \mathbb{R}^{p \times n}$ is given by

$$\mathbf{K}_k = (\mathbf{C} + \mathbf{B}^T \mathbf{P}_{k+1} \mathbf{B})^{-1} \mathbf{B}^T P_{k+1} \mathbf{M} \tag{5.3.23}$$

and the feedforward gain $\mathbf{G}_k \in \mathbb{R}^{p \times n}$ is given by

$$\mathbf{G}_k = (\mathbf{C} + \mathbf{B}^T \mathbf{P}_{k+1} \mathbf{B})^{-1} \mathbf{B}^T. \tag{5.3.24}$$

Since the matrices $\mathbf{P}_k$ are independent of observations, so are the feedback matrices $\mathbf{K}_k$ and hence can be precomputed, if need be. However, from the linear recursion (5.3.18) it is immediate that $\mathbf{g}_k$ depends on $\{\mathbf{z}_k, \mathbf{z}_{k+1}, \dots, \mathbf{z}_{N-1}\}$. Hence, this control sequence cannot be computed on-line.

5.3.6 Optimal Trajectory

Now substituting (5.3.22) into the model equation (5.3.1) and simplifying we get the evolution of the optimal trajectory given by

$$\mathbf{x}_{k+1} = (\mathbf{M} - \mathbf{B}\mathbf{K}_k)\mathbf{x}_k + \mathbf{B}\mathbf{G}_k\mathbf{g}_{k+1} \tag{5.3.25}$$

or

$$\mathbf{x}_{k+1} = \mathbf{M}\mathbf{x}_k - \mathbf{B}(\mathbf{K}_k\mathbf{x}_k - \mathbf{G}_k\mathbf{g}_{k+1}). \tag{5.3.26}$$

Computation of the optimal control and the trajectory is given in Algorithm 5.2.

5.3.7 Identification of Model Error

The second term on the right hand side of (5.3.26) is the optimal forcing term $\mathbf{B}\mathbf{u}_k$ and it plays a dual role. First, it forces the model trajectory toward the observations when the measure of closeness depends on the choice of the parameter p, the dimension of the optimal control vector $\mathbf{u}_k$, the matrix $\mathbf{B} \in \mathbb{R}^{n \times p}$ that helps to spread the effect of the force $\mathbf{u}_k$ to all the components of the state vector $\mathbf{x}_k$ and the matrix $\mathbf{C}$ which defines the energy norm of the control where it is assumed that the covariance matrix $\mathbf{R}$ of the observation noise is given. Consequently, $\mathbf{B}\mathbf{u}_k$ term contains the information about the model error.

Algorithm 5.2 PMP based algorithm for computing the optimal control and trajectory

---

Step 1: Given (5.3.1) and (5.3.2), compute $\mathbf{E} = \mathbf{B}\mathbf{C}^{-1}\mathbf{B}^T, \mathbf{F} = \mathbf{H}^T\mathbf{R}^{-1}\mathbf{H}$ and $\mathbf{W} = \mathbf{H}^T\mathbf{R}^{-1}$.

Step 2: Solve the nonlinear matrix Riccati equation (5.3.17) and compute the sequence P_k starting with $P_N = 0$. (This can be precomputed).

Step 3: Solve linear vector difference equation (5.3.18) backward starting with $g_N = 0$ since g_k depends on the observation, notice that the optimal control is a function of the observations.

Step 4: Knowing P_k, g_k we can compute the optimal control as

$$u_k = -\mathbf{C}^{-1}\mathbf{B}^T[P_{k+1}\mathbf{x}_{k+1} + g_{k+1}] \tag{5.3.27}$$

Step 5: The optimal trajectory is then given by (5.3.25) or (5.3.26).

5.4   Computation of Model Errors

In this section we extend the PMP one step further to compute the model error—errors due to incomplete physics, inappropriate parameterization and incorrect specification of initial conditions, and boundary conditions.

Let $\mathbf{S} \in \mathbb{R}^{n \times n}$ be a correction matrix to be added to $\mathbf{M}$ so that the new matrix $(\mathbf{M} + \mathbf{S})$ would represent the corrected model. If $\mathbf{x}_k$ is the optimal trajectory obtained from (5.3.25), then we require that

$$\mathbf{x}_{k+1} = \mathbf{M}\mathbf{x}_k + \mathbf{B}\mathbf{u}_k = (\mathbf{M} + \mathbf{S})\mathbf{x}_k. \tag{5.4.1}$$

This requirement forces the correction matrix $\mathbf{S}$ to satisfy the condition for $1 \leq k \leq N$,

$$\mathbf{S}\mathbf{x}_k = \mathbf{y}_k, \tag{5.4.2}$$

with $\mathbf{y}_k = \mathbf{B}\mathbf{u}_k$ and $\mathbf{u}_k$ the optimal sequence given in (5.3.22). Stated in other words, given the optimal state sequence $\{\mathbf{x}_k\}$ and the transformed optimal control sequence $\{\mathbf{y}_k = \mathbf{B}\mathbf{u}_k\}$, our goal is find a time invariant matrix $\mathbf{S} \in \mathbb{R}^{n \times n}$ such that (5.4.2) holds. That is, the matrix $\mathbf{S}$ is to be obtained by solving the linear inverse problem in (5.4.2).

To this end, we introduce a new quadratic cost function $\mathbf{Q} : \mathbb{R}^{n \times n} \rightarrow \mathbb{R}$ defined by

$$\mathbf{Q}(S) = \sum_{k=1}^{N} \mathbf{Q}_k(\mathbf{S}), \tag{5.4.3}$$

where

$$\mathbf{Q}_k(\mathbf{S}) = \frac{1}{2}(\mathbf{S}\mathbf{x}_k - \mathbf{y}_k)^T(\mathbf{S}\mathbf{x}_k - \mathbf{y}_k)$$

$$= \frac{1}{2}\left[\alpha\,(\mathbf{S}, \mathbf{x}_k) - 2\beta\,(\mathbf{S}, \mathbf{x}_k, \mathbf{y}_k) + \gamma\,(\mathbf{y}_k)\right], \tag{5.4.4}$$

with

$$\begin{cases} \alpha(\mathbf{S}, \mathbf{x}) &= \mathbf{x}^T(\mathbf{S}^T\mathbf{S})\mathbf{x}, \\ \beta(\mathbf{S}^*, \mathbf{y}) &= \mathbf{y}^T\mathbf{S}x, \\ \gamma(\mathbf{y}) &= \mathbf{y}^T\mathbf{y}. \end{cases} \tag{5.4.5}$$

The optimal correction matrix $\mathbf{S}$ is obtained as the minimizer of $\mathbf{Q}(\mathbf{S})$ in (5.4.3).

Recall that the gradient of $Q(S)$ with respect to S is a matrix given by

$$\nabla_S Q(S) = \left[\frac{\partial Q(S)}{\partial S_{ij}}\right] = \sum_{k=1}^{N}\left[\frac{\partial Q(S)}{\partial S_{ij}}\right], \qquad (5.4.6)$$

where

$$\left[\frac{\partial Q(S)}{\partial S_{ij}}\right] = \frac{1}{2}\left[\frac{\partial \alpha(S, x_k)}{\partial S_{ij}}\right] - \left[\frac{\partial \beta(S, x_k, y_k)}{\partial S_{ij}}\right]. \qquad (5.4.7)$$

5.4.1 Gradient of $\alpha(S, x)$

To compute the first term on the right hand side of (5.4.7), first express the matrix S in the row partitioned form

$$S = \begin{bmatrix} S_{1\star} \\ S_{2\star} \\ \vdots \\ S_{n\star} \end{bmatrix}, \qquad (5.4.8)$$

where $S_{i\star}$ is the ith row of $S \in \mathbb{R}^{n \times n}$. Then the Grammian matrix is given by

$$S^T S = \sum S_{i\star}^T S_{i\star}. \qquad (5.4.9)$$

Consequently, from (5.4.9)

$$\alpha(S, x) = x^T\left(S^T S\right)x = \sum_{i=1}^{N} x^T S_{i\star}^T S_{i\star} x = \sum_{i=1}^{N}(S_{i\star}x)^2 \qquad (5.4.10)$$

where $S_{i\star}x$ is the inner product of the ith row $S_{i\star}$ of S and the column vector x. Thus, the gradient of $\alpha(S, x)$ with respect to the column vector $S_{i\star}^T$ is given by

$$\nabla_{S_{i\star}^T}\alpha(S, x) = 2(xx^T)S_{i\star}^T. \qquad (5.4.11)$$

Taking the transpose of both sides, we get (since xx^T is symmetric)

$$\nabla_{S_{i\star}}\alpha(S, x) = 2S_{i\star}(xx^T). \qquad (5.4.12)$$

By stacking these rows of derivatives of $\alpha(S, x)$

$$\nabla_S\alpha(S, x) = 2S(xx^T). \qquad (5.4.13)$$

5.4.2   Gradient of $\beta(\mathbf{S}, \mathbf{x}, \mathbf{y})$

From (5.4.5),

$$\beta(\mathbf{S}, \mathbf{x}, \mathbf{y}) = \mathbf{y}^T \mathbf{S} \mathbf{x} = \sum_{i=1}^{N} \sum_{j=1}^{N} y_i S_{ij} x_j. \tag{5.4.14}$$

Hence, the matrix gradient of $\beta(\mathbf{S}, \mathbf{x}, \mathbf{y})$ is given by

$$\nabla_S \beta(\mathbf{S}, \mathbf{x}, \mathbf{y}) = [y_i x_j] = \mathbf{y} \mathbf{x}^T \in \mathbb{R}^{n \times n}. \tag{5.4.15}$$

5.4.3   Gradient of $\mathbf{Q}(\mathbf{S})$

Combining the expressions in (5.4.13)–(5.4.15) we obtain that

$$\nabla_S \mathbf{Q}(\mathbf{S}) = \mathbf{S} \left(\sum_{k=1}^{N} \mathbf{x}_k \mathbf{x}_k^T \right) - \left(\sum_{k=1}^{N} \mathbf{y}_k \mathbf{x}_k^T \right). \tag{5.4.16}$$

Hence ,the minimizer $\mathbf{S}$ is given by

$$\mathbf{S} = \left[\sum_{k=1}^{N} \mathbf{y}_k \mathbf{x}_k^T \right] \left[\sum_{k=1}^{N} \mathbf{x}_k \mathbf{x}_k^T \right]^+, \tag{5.4.17}$$

where $\mathbf{A}^+$ denotes the generalized inverse of $\mathbf{A}$.

5.5   An Application: Linear Advection Equation Model and Its Solution

In this section we illustrate the power of the Pontryagin's framework using the linear Burgers' equation which is essentially a linear advection equation in one dimension given by

$$\frac{\partial q}{\partial t} + \sin(x) \frac{\partial q}{\partial x} = 0, \ 0 \le x \le 2\pi. \tag{5.5.1}$$

with initial condition $q(x, 0) = \sin(x)$, and boundary condition $q(0, t) = q(2\pi, t) = 0$. We seek a solution to (5.5.1) by the method of characteristics (Carrier and Pearson 1976). The characteristics of (5.5.1) are given by

$$\frac{1 + \cos(x)}{1 - \cos(x)} e^{2t} = \frac{1 + \cos(x_0)}{1 - \cos(x_0)}, \tag{5.5.2}$$

where x_0 is the intersection of a particular characteristic curve with line of initial time ($t = 0$). Using the mathematical expression for the characteristic with the initial condition, it can be verified that the analytic solution is

$$q(x, t) = \frac{2 e^t \sin(x)}{1 + e^{2t} + \cos(x)(e^{2t} - 1)}. \tag{5.5.3}$$

Using this analytic solution, the profiles of the wave at times $t = 0, 0.5, 1.0, 1.5$ and 2.0 are shown in Fig. 5.1. The slope of the wave is finite at $x = \pi$ but approaches ∞ as $t \to \infty$.

Let $\tilde{q}_k(t)$ be the (exact) value of the kth Fourier coefficient of the solution $q(x, t)$ in (5.5.3) given by

$$\tilde{q}_k(t) = \frac{1}{\pi} \int_0^{2\pi} q(x, t) \sin(kx) dx. \tag{5.5.4}$$

Define the vector

$$\tilde{\mathbf{Q}}_t = (\tilde{q}_1(t), \tilde{q}_2(t), \dots, \tilde{q}_n(t))^T \in \mathbb{R}^n \tag{5.5.5}$$

of the first n Fourier coefficients of $q(x, t)$. The values of the coefficients $\tilde{q}_k(t)$ (computed using the well known quadrature formula 5.5.4) for $1 \le k \le n = 8$ and $0 \le t \le 2.0$ in steps of $\Delta t = 0.2$ are displayed in Table 5.1.

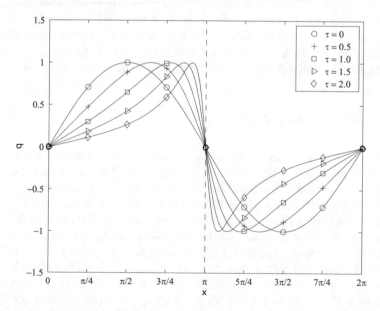

Fig. 5.1 A plot of the solution $q(x, t)$ at times $t = 0, 0.5, 1, 1.5, 2$

Table 5.1 Values of the first eight Fourier coefficients of $q(x,t)$ in (4.3) for various times computed using quadrature formula

t	$\tilde{q}_1(t)$	$\tilde{q}_2(t)$	$\tilde{q}_3(t)$	$\tilde{q}_4(t)$	$\tilde{q}_5(t)$	$\tilde{q}_6(t)$	$\tilde{q}_7(t)$	$\tilde{q}_8(t)$
0.0	1.000	0.000	0.000	0.000	0.000	0.000	0.000	0.000
0.2	0.990	−0.099	0.010	−0.001	0.000	−0.000	0.000	−0.000
0.4	0.961	−0.190	0.037	−0.007	0.001	−0.000	0.000	−0.000
0.6	0.915	−0.267	0.078	−0.023	0.007	−0.002	0.001	−0.000
0.8	0.856	−0.325	0.124	−0.047	0.018	−0.007	0.003	−0.001
1.0	0.786	−0.363	0.168	−0.078	0.036	−0.017	0.008	−0.004
1.2	0.712	−0.382	0.205	−0.110	0.059	−0.032	0.017	−0.009
1.4	0.634	−0.384	0.232	−0.140	0.085	−0.051	0.031	−0.019
1.6	0.559	−0.371	0.247	−0.164	0.109	−0.072	0.048	−0.032
1.8	0.487	−0.349	0.250	−0.179	0.128	−0.092	0.066	−0.047
2.0	0.420	−0.320	0.244	−0.153	0.141	−0.108	0.082	−0.062

An n-mode approximation (resulting from spectral truncation to $q(x,t)$) is then given by

$$\tilde{q}(x,t) \stackrel{\circ}{=} \sum_{k=1}^{n} \tilde{q}_k(t) \sin(kx). \tag{5.5.6}$$

Let us denote the 4-mode approximation by $\tilde{q}_1(x,t)$ and the 8-mode approximation by $\tilde{q}_2(x,t)$.

A comparison of the exact solution $q(x,t)$ with the 4-mode approximation $\tilde{q}_1(x,t)$ and the 8-mode approximation $\tilde{q}_2(x,t)$ obtained from (5.5.6) with $n = 4$ and 8, respectively, at $t = 2.0$ is given in Fig. 5.2. Clearly, the 8-mode approximation $\tilde{q}_2(x,t)$ is closer to the true solution $q(x,t)$ than the 4-mode approximation $\tilde{q}_1(x,t)$. Further, the errors are the greatest at the point of extreme steepness of waves.

5.5.1 The Low-Order Model (LOM)

In demonstrating the power of Pontryagin's method developed in this chapter our first step is to obtain discrete time model equations. There are at least two ways, in principle, to achieve this goal. The first way is to directly discretize (5.5.1) by embedding a grid in the two dimensional domain with $0 \le x \le 2\pi$ and $t \ge 0$. The second way is to project the infinite dimensional system (5.5.1) onto a finite dimensional space using the standard Galerkin projection method and obtain a system of n ordinary differential equations (ODE's) describing the evolution of the Fourier amplitudes $\tilde{q}_i(t)$ in (5.5.1), $1 \le i \le n$. The resulting nth order system is known as the low-order model (LOM). Then LOM can be discretized using one of several known methods. In this chapter we use this latter spectral approach using LOM. The Fourier series of $q(x,t)$ consists of an infinite series of sine waves given by

$$q(x,t) = \sum_{n=1}^{\infty} [q_n(t) \sin(nx)]. \tag{5.5.7}$$

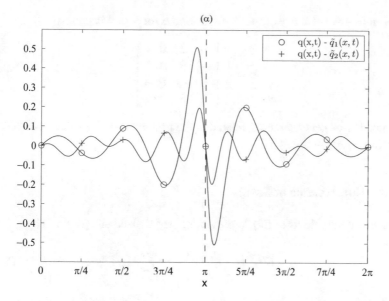

Fig. 5.2 Comparison of the error $q(x, t) - \tilde{q}_1(x, t)$ in the 4-mode approximation $\tilde{q}(x, t) = \sum_{i=1}^{4} \tilde{q}_i(t) \sin(ix)$ and the error $q(x, t) - \tilde{q}_2(x, t)$ in the 8-mode approximation $\tilde{q}_2(x, t) = \sum_{i=1}^{8} \tilde{q}_i(t) \sin(ix)$ at $t = 2.0$

An LOM (n) describing evolution of the amplitudes of the spectral components is obtained by exploiting the orthogonality properties of the $\sin(kx)$ functions for $1 \leq k \leq n$. Substituting (5.5.7) in (5.5.1), multiplying both sides by $\sin(kx)$ and integrating both sides from 0 to 2π we obtain the LOM (n). This approach follows Platzman (1964). In matrix form, the governing dynamics is given by

$$\frac{d\mathbf{q}(t)}{dt} = \mathbf{A}\mathbf{q}(t), \tag{5.5.8}$$

where

$$\mathbf{q}(t) = (q_1(t), q_2(t), q_n(t))^T, \text{ and } \mathbf{q}(0) = (1, 0, 0, \ldots, 0)^T, \tag{5.5.9}$$

as its initial condition. The matrix $\mathbf{A}$ is given by

$$\mathbf{A} = \frac{1}{2} \begin{bmatrix} 0 & c_1 & 0 & 0 & \cdots & 0 & 0 & 0 \\ a_2 & 0 & c_2 & 0 & \cdots & 0 & 0 & 0 \\ 0 & a_3 & 0 & c_3 & \cdots & 0 & 0 & 0 \\ \cdot & \cdot & \cdot & \cdot & & \cdot & \cdot & \cdot \\ \cdot & \cdot & \cdot & \cdot & & \cdot & \cdot & \cdot \\ \cdot & \cdot & \cdot & \cdot & & \cdot & \cdot & \cdot \\ 0 & 0 & 0 & 0 & \cdots a_{n-1} & 0 & c_{n-1} \\ 0 & 0 & 0 & 0 & \cdots & 0 & a_n & 0 \end{bmatrix} \tag{5.5.10}$$

where $a_i = -(i-1)$, $c_i = (i+1)$. An example for $n = 4$ is given by

$$
\mathbf{A} = \frac{1}{2} \begin{bmatrix} 0 & 2 & 0 & 0 \\ -1 & 0 & 3 & 0 \\ 0 & -2 & 0 & 4 \\ 0 & 0 & -3 & 0 \end{bmatrix}. \tag{5.5.11}
$$

A number of generalizations related to the properties of the solution of the LOM (n) in (5.5.8) are in order:

5.5.1.1   Conservation of Energy

Consider a quadratic form $E(q)$ representing generalized energy given by

$$
E(\mathbf{q}) = \frac{1}{2} \mathbf{q}^T \mathbf{K} \mathbf{q} = \frac{1}{2} \sum_{k=1}^{n} K_k q_k^2 \tag{5.5.12}
$$

where

$$
\mathbf{K} = \mathrm{Diag}(1, 2, 3, \ldots, i, \ldots, n) \tag{5.5.13}
$$

is a diagonal matrix with the indicated entries as its diagonal elements. It can be verified that the time derivative of $E(q)$ evaluated along the solution of (5.5.8) is given by

$$
\frac{dE(\mathbf{q})}{dt} = \mathbf{q}^T \mathbf{K} \frac{d\mathbf{q}}{dt} = \mathbf{q}^T \mathbf{K} \mathbf{A} \mathbf{q}. \tag{5.5.14}
$$

Using the form of $\mathbf{K}$ in (5.5.13) and $\mathbf{A}$ in (5.5.10), it is an easy exercise to verify that the product $\mathbf{K}\mathbf{A}$ is a skew-symmetric matrix given by

$$
\mathbf{K}\mathbf{A} = \frac{1}{2} \begin{bmatrix}
0 & s_1 & 0 & 0 & \cdots & 0 & 0 & 0 \\
-s_1 & 0 & s_2 & 0 & \cdots & 0 & 0 & 0 \\
0 & -s_2 & 0 & s_3 & \cdots & 0 & 0 & 0 \\
\cdot & \cdot & \cdot & \cdot & & \cdot & \cdot & \cdot \\
\cdot & \cdot & \cdot & \cdot & & \cdot & \cdot & \cdot \\
\cdot & \cdot & \cdot & \cdot & & \cdot & \cdot & \cdot \\
0 & 0 & 0 & 0 & \cdots & -s_{n-2} & 0 & s_{n-1} \\
0 & 0 & 0 & 0 & \cdots & 0 & -s_{n-1} & 0
\end{bmatrix} \tag{5.5.15}
$$

where $s_i = i(i+1)$ for $1 \le i \le n-1$. Hence, the quadratic form $\mathbf{q}^T \mathbf{K} \mathbf{A} \mathbf{q}$ is zero which in turn implies that the energy $E(q)$ is conserved by (5.5.8), and

$$
\frac{dE(\mathbf{q})}{dt} = 0. \tag{5.5.16}
$$

A consequence of (5.5.16) is that the solution $\mathbf{q}(t)$ of (5.5.8) lies on the surface of an n-dimensional ellipsoid defined by

$$\sum_{k=1}^{n} \mathbf{K}_k \mathbf{q}_k^2(t) = \sum_{k=1}^{n} \mathbf{K}_k \mathbf{q}_k^2(0) = 1. \tag{5.5.17}$$

Clearly, the length of the kth semi-axis of this ellipsoid is given by $(1/k)^{1/2}$. Hence the volume of this ellipsoid is given by

$$\text{Volume} = \frac{4}{3}\pi(\frac{1}{n!})^{1/2} \tag{5.5.18}$$

Since $n! = \mathcal{O}(2^{n\log n})$, it turns out that the volume of this ellipsoid goes to zero at an exponential rate as n increases signaling degeneracy for large n.

5.5.2   Solution of LOM(n) in (5.5.8)

It turns out that quite like the PDE (5.5.1), its LOM(n) counterpart in (5.5.8) can be solved exactly. The process of obtaining its solution is quite involved. To minimize the digression from the main development, we have chosen to describe this solution process in the Appendix of this chapter. The eigenstructure of $\mathbf{A}$, its Jordan canonical form and the form of the general solution of (5.5.6) are discussed in detail in the Appendix. Our goal is to use the exact solution of (5.5.8) given in Appendix "Solution of the LOM in (5.5.8)" to calibrate the choice of Δt, the time discretization interval to be used in the following section.

Numerical Experiments
Discretizing the spectral model (5.5.8) with $n = 4$ using the first-order Euler scheme, we obtain

$$\boldsymbol{\xi}_{k+1} = \mathbf{M}\boldsymbol{\xi}_k \tag{5.5.19}$$

where $\boldsymbol{\xi}_k = \mathbf{q}(t = k\Delta t)$ and Δt denotes the length of the time interval used in time discretization and

$$\mathbf{M} = (\mathbf{I} + \Delta t\mathbf{A}) \in \mathbb{R}^{n \times n}, \tag{5.5.20}$$

where $\mathbf{A} \in \mathbb{R}^{4 \times 4}$ is given in (5.5.11) and the initial condition in (5.5.9). PMP approach requires the addition of the forcing term to (5.5.1). The forced version of (5.5.1) is then represented as

$$\mathbf{x}_{k+1} = \mathbf{M}\mathbf{x}_k + \mathbf{B}\mathbf{u}_k, \quad \mathbf{x}_0 = \mathbf{q}(0), \tag{5.5.21}$$

where $\mathbf{u}_k \in \mathbb{R}^p$ and $\mathbf{B} \in \mathbb{R}^{n \times p}$. Equation (5.5.21) defines the evolution of the spectral amplitudes. Compared to the original equation, the spectral model in (5.5.21) has two types of model errors: first from the spectral truncation in the Galerkin projection and the error due to finite differencing (5.5.8) using the first-order method.

Observations We propose to use the exact Fourier coefficient vector $\tilde{\mathbf{Q}}_k = \tilde{\mathbf{Q}}_t$ at $t = k\Delta t$ in (5.5.5) corrupted by additive noise as the observations in our numerical experiments. Let

$$\mathbf{z}_k = \tilde{\mathbf{Q}}_k + \mathbf{v}_k, \tag{5.5.22}$$

where $\mathbf{z}_k \in \mathbb{R}^n, \tilde{\mathbf{Q}}_k \in \mathbb{R}^n, \mathbf{v}_k \sim N(0, \mathbf{R})$ and $\mathbf{R} = \sigma_0^2 \mathbf{I}_n$.

The form of the functional V_k is given by

$$V_k = \frac{1}{2}(\mathbf{z}_k - \mathbf{x}_k)^T \mathbf{R}^{-1}(\mathbf{z}_k - \mathbf{x}_k) + \frac{1}{2}\mathbf{u}_k^T \mathbf{C}\mathbf{u}_k \tag{5.5.23}$$

where $\mathbf{C} \in \mathbb{R}^{p \times p}$ is a symmetric and positive definite matrix.

Applying the results from Sect. 5.3, it follows that

$$\mathbf{u}_k = -\mathbf{C}^{-1}\mathbf{B}^T \lambda_{k+1}, \tag{5.5.24}$$

where

$$\lambda_k = \mathbf{M}^T \lambda_{k+1} + \mathbf{R}^{-1}(\mathbf{x}_k - \mathbf{z}_k), \tag{5.5.25}$$

with $\lambda_N = 0$.

The TPBVP problem in (5.3) and (5.7) is then solved using the method described in Sect. 5.3. Accordingly,

$$\lambda_k = \mathbf{P}_k \mathbf{x}_k - \mathbf{g}_k, \tag{5.5.26}$$

where

$$\mathbf{P}_k = \mathbf{M}^T \mathbf{P}_{k+1}(\mathbf{I} + \mathbf{E}\mathbf{P}_{k+1})^{-1}\mathbf{M} + \mathbf{F}, \tag{5.5.27}$$

$$\mathbf{g}_k = \mathbf{M}^T \mathbf{g}_{k+1} - \mathbf{M}^T \mathbf{P}_{k+1}(\mathbf{I} + \mathbf{E}\mathbf{P}_{k+1})^{-1} \mathbf{E}\mathbf{g}_{k+1} + \mathbf{W}\mathbf{z}_k \tag{5.5.28}$$

where $\mathbf{E} = \mathbf{B}\mathbf{C}^{-1}\mathbf{B}^T, \mathbf{F} = \mathbf{R}^{-1}, \mathbf{W} = R^{-1}, \mathbf{P}_N = 0$, and $\mathbf{g}_N = 0$. Solving (5.5.27)–(5.5.28), we then assemble $\mathbf{u}_k$ using (5.5.24). Substituting it in (5.5.21) we get the optimal solution.

Experiment 5.1. In this first experiment, we set $n = m = p = 4, \mathbf{B} = \mathbf{I}_4$ and $\mathbf{u}_k \in \mathbb{R}^4$. That is, the number of observations, the dimension of external forcing, and level of spectral truncation are all 4. The uncontrolled model is

$$\xi_{k+1} = \mathbf{M}\xi_k \tag{5.5.29}$$

and the controlled model is

$$\mathbf{X}_{k+1} = \mathbf{M}\mathbf{X}_k + \mathbf{B}\mathbf{u}_k \tag{5.5.30}$$

with $\mathbf{M} = (\mathbf{I} + \Delta t \mathbf{A})$ and $\mathbf{A}$ is given in (5.5.11).

Both the models start from the same initial condition $\boldsymbol{\xi}_0 = \mathbf{x}_0 = (1.1, 0, 0, 0)^T$ which is different from the one that was used to generate the observations. The model inherits three types of errors: (1) spectral truncation, (2) finite differencing, and (3) initial condition uncertainty. The power of PMP is that the optimal control term $\mathbf{B}\mathbf{u}_k$ compensates for all three types of errors.

The observation vector $\mathbf{z}_k \in \mathbb{R}^4$ is given by

$$\mathbf{z}_k = \tilde{\mathbf{Q}}_k + \boldsymbol{v}_k \tag{5.5.31}$$

for $1 \le k \le 10$ where $\tilde{\mathbf{Q}}_k = (\tilde{q}_1(k), \tilde{q}_2(k), \tilde{q}_3(k), \tilde{q}_4(k))^T$ given in Table 5.1, and $v_k \sim N(0, \mathbf{R})$.

It is further assumed that $\mathbf{R} = \sigma_0^2 \mathbf{I}_n$ and $\mathbf{C} = c\mathbf{I}_p$. Substituting these in the expression for V_k, it can be verified that

$$V_k = \frac{1}{2}\sigma_0^2 \langle \mathbf{z}_k - \mathbf{H}\mathbf{x}_k, \mathbf{z}_k - \mathbf{H}\mathbf{x}_k \rangle + c\langle \mathbf{u}_k, \mathbf{u}_k \rangle. \tag{5.5.32}$$

A comparison of the evolution of the four components of the uncontrolled error, $\mathbf{e}_0 = \boldsymbol{\xi}_k - \mathbf{z}_k \in \mathbb{R}^4$ and the corresponding components of the controlled error, $\mathbf{e}_c = \mathbf{x}_k - \mathbf{z}_k \in \mathbb{R}^4$ when $\sigma^2 = 0.001$ fixed but c is varied through $10^5, 10^3$ and 1 are given in the four panels (a) to (d) of Fig. 5.3. It is clear from this Fig. 5.3 that the magnitudes of the individual components of the controlled error are uniformly (in time k) less than the magnitudes of the corresponding components of the uncontrolled error. Further, the magnitudes of the controlled error decrease with the decrease in the value of the control parameter c.

This behavior can be easily explained from (5.5.32). When the value of the control parameter c is large (for a fixed $\mathbf{R}^{-1}$), the minimization process forces $\mathbf{u}_k$ to be small. However, if c is small, the minimization allows for larger value of $\mathbf{u}_k$. This increased forcing helps to move $\mathbf{x}_k$ in such a way that $\mathbf{H}\mathbf{x}_k$ is closer to $\mathbf{z}_k$. This observation is corroborated by the plots of the evolution of the four components of the control $\{\mathbf{u}_k\}$ given in the four panels (a) to (d) in Fig. 5.4. It is evident from this Fig. 5.4 that the magnitude of the initial values of the control increases as the parameter c is decreased.

A standard measure of the closeness of the jth component of the controlled and uncontrolled model solution with the jth component of the observations are given by

$$RMS_{1j} = \left[\frac{1}{N}\sum_{k=0}^{N}(z_{k,j} - x_{k,j})^2\right]^{1/2} \tag{5.5.33}$$

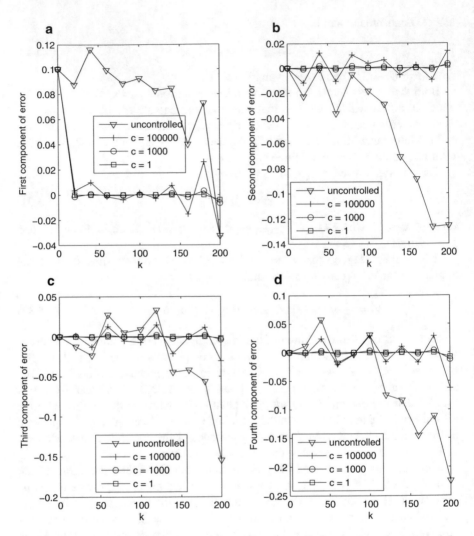

Fig. 5.3 Comparison of the four components of the uncontrolled error $e_0 = \xi_k - z_k$, and the controlled error $e_c = \mathbf{x}_k - \mathbf{z}_k$ for $p = 4$, $\mathbf{B} = \mathbf{I}_4$, $c = \{100{,}000, 1000, 1\}$. (**a**) First component of error, $\sigma_0^2 = 0.001$. (**b**) Second component of error, $\sigma_0^2 = 0.001$. (**c**) Third component of error, $\sigma_0^2 = 0.001$. (**d**) Fourth component of error, $\sigma_0^2 = 0.001$

and

$$RMS_{2j} = \left[\frac{1}{N} \sum_{k=0}^{N} (z_{k,j} - \xi_{k,j})^2 \right]^{1/2}. \tag{5.5.34}$$

Table 5.2 gives the values of these measure for various combinations of the values of σ_0^2 and c. It is clear from Fig. 5.3 and Table 5.2 that for a given σ_0^2, RMS_1 decreases as c decreases.

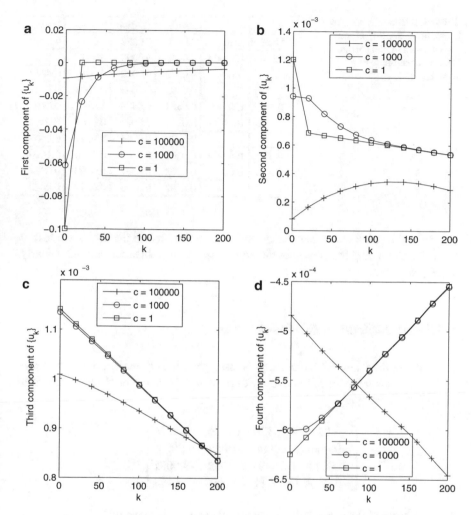

Fig. 5.4 Comparison of four components of the control sequence uk for $p = 4, \mathbf{B} = \mathbf{I}_4$, and $c = \{100,000, 1000, 1\}$. (**a**) First component of $\{\mathbf{u}_k\}, \sigma_0^2 = 0.001$. (**b**) Second component of $\{\mathbf{u}_k\}, \sigma_0^2 = 0.001$. (**c**) Third component of $\{\mathbf{u}_k\}, \sigma_0^2 = 0.001$. (**d**) Fourth component of $\{\mathbf{u}_k\}, \sigma_0^2 = 0.001$

Experiment 5.2. In this experiment we set $p = 1$ (dimension of forcing) and $\mathbf{B} = (1111)^T$ and all the other parameters are the same as in Experiment 1. A comparison of the plots of the observations with controlled and uncontrolled model solution is given in Fig. 5.5. Table 5.3 provides a comparison of the RMS errors for various choices of σ_0^2 and c. Recall that when $p = 1$, the same control is applied to every component of the state vector as opposed to when $p = 4$ where different elements of the control vector impact the evolution of the different components of the state vector. Thus in Experiment 1 ($p = 4$) the components of the control vector are

Table 5.2 Root mean square errors of the controlled and uncontrolled model solution with observations $p = 4$, $\mathbf{B} = \mathbf{I}_4$.

σ_0^2	c	RMS_{e1}	RMS_{e2}	RMS_{e3}	RMS_{e4}
Uncontrolled		0.0850	0.0658	0.0551	0.0958
0.001	100,000	0.0333	0.0090	0.0140	0.0258
	1000	0.0302	0.0013	0.0016	0.0042
	1	0.0302	0.0006	0.0006	0.0023
0.005	100,000	0.0654	0.0438	0.0345	0.0465
	1000	0.0323	0.0084	0.0061	0.0112
	1	0.0302	0.0005	0.0007	0.0027
0.01	100,000	0.0901	0.1160	0.1200	0.1010
	1000	0.0368	0.0337	0.0360	0.0284
	1	0.0302	0.0005	0.0072	0.0028

customized to each component of the state vector and hence the errors are less as borne by comparing the corresponding elements of the Tables 5.2 and 5.3. Clearly, larger p is better.

5.5.3   Identification of Model Error

Using the optimal control sequence $\mathbf{y}_k = \mathbf{B}\mathbf{u}_k$ and its associated optimal trajectory $\mathbf{x}_k$ found in Experiment 5.2 above (with $n = 4$), the value of $\mathbf{S}$ computed from (5.4.17) is given by

$$\mathbf{S} = \begin{bmatrix} -0.0045 & -0.0176 & -0.0186 & -0.0074 \\ -0.0004 & -0.0006 & 0.0063 & 0.0084 \\ 0.0009 & 0.0043 & 0.0009 & -0.0081 \\ -0.0009 & -0.0011 & -0.0001 & -0.0173 \end{bmatrix} \qquad (5.5.35)$$

The trajectory of the corrected but uncontrolled model is given by

$$\boldsymbol{\zeta}_{k+1} = (\mathbf{M} + \mathbf{S})\boldsymbol{\zeta}_k, \quad \boldsymbol{\zeta}_0 = x_0. \qquad (5.5.36)$$

A comparison of $\boldsymbol{\zeta}_k - \mathbf{z}_k$, the error between corrected but uncontrolled model in (5.5.36), and $\boldsymbol{\xi}_k - \mathbf{z}_k$, the error between the uncorrected and uncontrolled model in (5.5.29) is given in Fig. 5.6. It is evident from this Fig. 5.6 that the trajectory of the corrected but uncontrolled model fits the observations better.

We conclude this section with the following remarks:

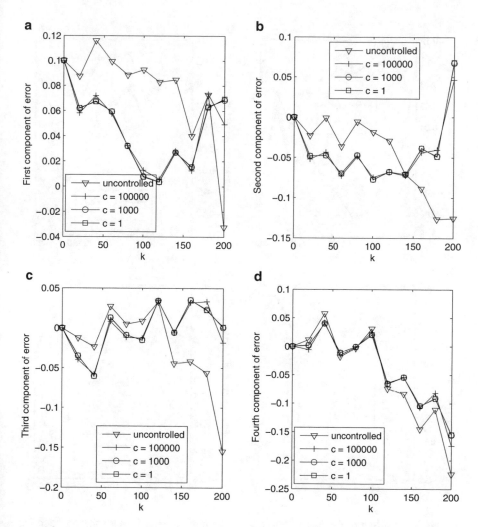

Fig. 5.5 Comparison of the four components of the uncontrolled error $e_0 = \xi_k - z_k$, and the controlled error $e_c = x_k - z_k$ for $p = 1, B = (1, 1, 1, 1)^T, c = \{100,000, 100, 1\}$. (**a**) First component of error, $\sigma_0^2 = 0.001$. (**b**) Second component of error, $\sigma_0^2 = 0.001$. (**c**) Third component of error, $\sigma_0^2 = 0.001$. (**d**) Fourth component of error, $\sigma_0^2 = 0.001$

1. Define a vector $\mathbf{s}(x) = (\sin x, \sin 2x, \sin 3x, \sin 4x)^T$ and define

$$q_1(x, k) = \xi_k^T \mathbf{s}(x)$$
$$q_2(x, k) = \zeta_k^T \mathbf{s}(x)$$

(5.5.37)

where ξ_k is the (uncontrolled) model trajectory obtained from (5.5.29) using matrix $\mathbf{M}$ and ζ_k is the (uncontrolled) model trajectory obtained from (5.5.36)

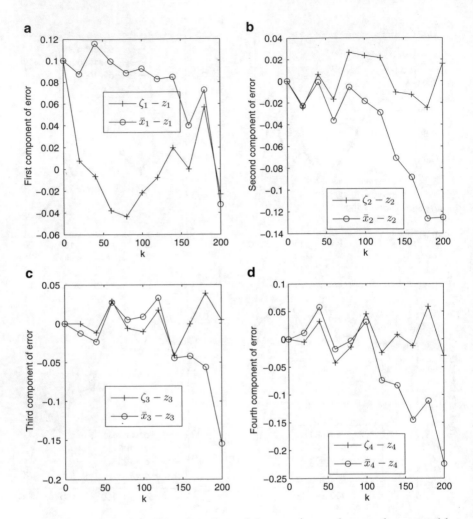

Fig. 5.6 Comparison of the four components of the error $\zeta_k - z_k$ between the corrected but uncontrolled model state $\{\zeta_k\}$ with the observations $\{z_k\}$ and the error $\bar{x}_k - z_k$ between the original uncorrected and uncontrolled model state $\{\bar{x}_k\}$ with the observations $\{z_k\}$. (**a**) First component of error, $\sigma_0^2 = 0.001$. (**b**) Second component of error, $\sigma_0^2 = 0.001$. (**c**) Third component of error, $\sigma_0^2 = 0.001$. (**d**) Fourth component of error, $\sigma_0^2 = 0.001$

with matrix $(\mathbf{M} + \mathbf{S})$. Clearly $\tilde{q}_1(x, k)$ and $\tilde{q}_2(x, k)$ are approximations to the exact solution $q(x, t)$ in (5.5.3) at $t = k\Delta t$. It can be verified that

$$|q(x, k) - \tilde{q}_2(x, k)| \leq |q(x, k) - \tilde{q}_1(x, k)|, \qquad (5.5.38)$$

where $q(x, k) = q(x, t)$ at $t = k\Delta t$. That is, the one-step model error correlation matrix $\mathbf{S}$ forces the model solution closer to the true solution.

Table 5.3 Root mean square errors of the controlled and uncontrolled model solution with observations $p = 1$, $\mathbf{B} = \mathbf{I}_4$.

σ_0^2	c	RMS_{e1}	RMS_{e2}	RMS_{e3}	RMS_{e4}
Uncontrolled		0.0850	0.0658	0.0551	0.0958
0.001	100000	0.0539	0.0551	0.0286	0.0728
	1000	0.0546	0.0568	0.0275	0.0693
	1	0.0546	0.0570	0.0275	0.0691
0.005	100000	0.0933	0.0869	0.0579	0.1010
	1000	0.0968	0.0821	0.0573	0.0962
	1	0.0973	0.0815	0.0573	0.0958
0.01	100000	0.1000	0.1330	0.1410	0.1560
	1000	0.0846	0.1140	0.1430	0.1490
	1	0.0834	0.1110	0.1450	0.1490

2. Only for simplicity in exposition did we pose the inverse problem in (5.4.2) as an unconstrained problem. In fact, one could readily accommodate structural constraint on $\mathbf{S}$ such as requiring it to be a diagonal, tridiagonal or lower-triangular matrix, etc. Further, we could also readily impose inequality constraints on a selected subset of elements of $\mathbf{S}$.

5.6 Exercises

5.6.1 Demonstrations

Demonstration: Derber's Problem

An interesting approach to correcting inherent model error due to incorrect dynamics was investigated by John Derber (Derber 1989: A variational continuous assimilation technique. Mon Weather Rev 117(11):2437–2446). The basic idea behind the work was that a barotropic model was assumed as the constraint for large-scale hemispheric waves in mid-latitude. However, the "true" model was a baroclinic model that accounted for thermodynamic processes that were absent from the barotropic model. Then, render the assumption that "observations" were obtained from the baroclinic models' evolution, find a correction term in the barotropic equations, correction in the space dimension only, such that the sum of squared differences between the augmented barotropic model and observations is minimized.

Let us set up the problem in 1-space dimension (x-coordinate) and time t. Linear advection $u(x, t)$ satisfies

$$\frac{\partial u}{\partial t} + c\frac{\partial u}{\partial x} = \phi(x),\ c = \text{ const.} > 0 \tag{DP.1}$$

Discretize

$$u_i^{n+1} - u_i^n = \frac{c\tau}{\Delta}\left(u_i^n - u_{i-1}^n\right) + \phi_i, \tag{DP.2}$$

where $u_i^n = n(i\Delta, n\tau)$. We choose $c\tau/\Delta = 1$ (Courant-Friedrich-Lewy (CFL) condition). (This scheme amplifies the solution and truncation error is significant with $c\tau/\Delta = 1$, but we are demonstrating the mechanics of the method and leave it to the reader to solve the problem with more accurate numerical scheme such as two-step leap-frog, Lax-Wendroff, or implicit scheme (Richtmyer and Morton 1967).

Under the CFL condition, the constraint becomes

$$u_i^{n+1} = u_{i-1}^n + \phi_i. \tag{DP.3}$$

Assume observations exists, $\tilde{u}_i^n$. Minimize the functional

$$J = \sum_{i,n}(u_i^n - \tilde{u}_i^n)^2, \tag{DP.4}$$

subject to the constraint

$$u_i^{n+1} = u_{i-1}^n + \phi_i. \tag{DP.5}$$

Assume a periodic solution over 8Δ spatial points $i = 0, 1, 2 \dots 7$ where $u_0^n = u_8^n, u_{-1}^n = u_7^n$, etc. If we have observations at all 8 points $i = 0, 1, 2 \dots 7$ at two time levels $n = 0, 1$, then we have constraints

$$
\begin{aligned}
u_0^1 &= u_7^0 + \phi_0 & u_0^2 &= u_7^1 + \phi_0 \\
u_1^1 &= u_0^0 + \phi_1 & u_1^2 &= u_0^1 + \phi_1 \\
&\cdots & &\cdots \\
u_7^1 &= u_6^0 + \phi_7 & u_7^2 &= u_6^1 + \phi_7.
\end{aligned}
\tag{DP.6}
$$

With forcing function $\phi(x)$ the only control, the first variation of J set to zero gives

$$\delta J = 0 = [(u_0^1 - \tilde{u}_0^1) + (u_0^2 - \tilde{u}_0^2) + (u_1^2 - \tilde{u}_1^2)] \, \delta\phi_0$$
$$+ [(u_1^1 - \tilde{u}_1^1) + (u_1^2 - \tilde{u}_1^2) + (u_2^2 - \tilde{u}_2^2)] \, \delta\phi_1$$
$$\cdots \tag{DP.7}$$
$$+ [(u_7^1 - \tilde{u}_7^1) + (u_7^2 - \tilde{u}_7^2) + (u_0^2 - \tilde{u}_0^2)] \, \delta\phi_7.$$

Therefore, the Euler-Lagrange equations are

$$\delta\phi_0 : u_0^1 - \tilde{u}_0^1 + u_0^2 - \tilde{u}_0^2 + u_1^2 - \tilde{u}_1^2 = 0$$

or since

$$u_0^1 = u_7^0 + \phi_0 \tag{DP.8}$$
$$u_0^2 = u_7^1 + \phi_0 = u_6^0 + \phi_7 + \phi_0$$
$$u_1^2 = u_0^1 + \phi_1 = u_7^0 + \phi_0 + \phi_1,$$

we get

$$3\phi_0 + \phi_1 + \phi_7 = \underbrace{2u_7^0 + u_6^0 - (\tilde{u}_0^1 + \tilde{u}_0^2 + \tilde{u}_1^2)}_{\text{known}} \tag{DP.9}$$

and similarly for the terms that multiply $\delta\phi_1, \delta\phi_2, \ldots, \delta\phi_7$. Thus, we get 8 equations in the eight unknowns $\phi_1, \phi_2, \ldots, \phi_7$. And so, by Derber's method, the missing term in the dynamical law, approximated by $\phi(x)$ can be found, and when added to the linear advection constraint, minimizes the functional J. A rather severe assumption in this approach is that the "missing term" in the dynamical law is considered to be a function of the space variable alone. DeMaria and Jones (1993) used Derber's method to determine a forcing term in the model used to forecast the track of Hurricane Hugo (1989). Results were encouraging for Hugo although marginal results were associated with application to four other hurricanes that occurred in 1989.

Outline the solution to this problem by using FSM and Pontryagin's minimum principle (PMP).

5.7   Notes and References

The subject of feedback control came into prominence shortly after World War
II (Bennett 1996; Bryson Jr 1996; Wiener 1961). Development of algorithms to
optimally track rockets and artificial satellites and effectively change their course
took center stage in the development of optimal control theory. On the basic
approach developed in this domain is the Pontryagin's minimum principle (PMP)
(Boltânski 1978; Boltânski et al. 1971; Pontryagin et al. 1962). Also refer to Athans
and Falb (1966), Bryson (1975), Lewis and Syrmos (1995) and Naidu (2002) for
more details. In essence, PMP embodies the search for a minimum under dynamic
constraints. It turns out the well known 4D-VAR is a special case of PMP.

   Within the content of dynamic data assimilation, PMP is used to force the model
solution towards the observation in much the same spirit as the nudging method
(Anthes 1974). Refer to Lakshmivarahan and Lewis (2013) for a critical review of
nudging methods.

   By introducing an appropriate Hamiltonian function, PMP reduces the well
known Euler-Lagrange equation to a system of two first order canonical Hamiltonian equations. A similar formulation has guided numerous developments in physics
(Goldstein 1950).

   While Kuhn (1982) extended the Lagrangian technique for equality constraints to
include inequality constraints by developing the theory of nonlinear programming
for static problems, Pontryagin et al. (1962) used the Hamiltonian formulation to
extend the Euler-Lagrange formulation in the calculus of variations. This extension
has been the basis for fundamental developments in the theory of optimal control in
the dynamic setting. The resulting framework is general enough to include both
equality and inequality constraints on both the state and the control. The close
relationship between the PMP and the Kuhn-Tucker condition is explored in Canon
(1970).

   While our approach to dynamic data assimilation and model error correction
using PMP is new, there has been considerable interest in the analysis of model error.
The literature in this vast area of model error correction may be broadly divided
along two dimensions—deterministic vs stochastic models and the use of model as
a strong or a weak constraint.

   Derber (1989) first postulates that the deterministic model error can be expressed
as a product of known time varying function and a fixed but unknown spatial
function Φ. Using the model as a strong constraint, he then extends the 4D-VAR
to estimate Φ along with the initial condition. To our knowledge this represents the
first attempt to use the variational framework to estimate the model error. Griffith
and Nichols (2000) postulate that the model error evolves according to an auxiliary
dynamical model with unknown parameters. By augmenting the given model with
this empirical auxiliary model, they the use 4D-VAR approach to estimate both
the initial condition and the parameters of the auxiliary model. The PMP based
approach described in this Chapter does not rely on any such empirical assumption.

Other approaches to estimating model errors in the content of control theory and meteorology include Rauch et al. (1965), Friedland (1969), Bennett and Thorburn (1992), Bennett (1992), and Dee and Da Silva (1998). In a similar vain, Ménard and Daley (1996) made the first attempt to relate Kalman smoother with PMP. But our development in this Chapter differs from that of Ménard and Daley (1996) in that while we use the deterministic model with time varying errors as a strong constraint, they use a stochastic model with random model error with known covariance structure as a weak constraint. Zupanski (1997) provides an elegant discussion of the advantages of the weak constraint formulation of the 4D-VAR to assess the impact of systematic and random errors which is many ways compliments the developments in Ménard and Daley (1996).

Appendix

Solution of the LOM in (5.5.8)

In this appendix we analyze the eigenstructure of the matrix A in (5.5.8).

Eigenstructure of the Matrix **A**

Since the structure of the matrix **A** in (5.5.8) is closely related to the tridiagonal matrix, we start this discussion by stating a well known result relating to the recursive computation of the determinant of the tridiagonal matrix (Lakshmivarahan and Dhall 1990, Chap. 6).

Let $\mathbf{B}_k \in \mathbb{R}^{k \times k}$ be a general tridiagonal matrix of the form

$$
\mathbf{B}_k = \begin{bmatrix}
b_1 & c_1 & 0 & 0 & \cdot & \cdot & \cdot & 0 & 0 & 0 \\
a_2 & b_2 & c_2 & 0 & \cdot & \cdot & \cdot & 0 & 0 & 0 \\
\cdot & \cdot & \cdot & \cdot & & & & \cdot & \cdot & \cdot \\
\cdot & \cdot & \cdot & \cdot & & & & \cdot & \cdot & \cdot \\
\cdot & \cdot & \cdot & \cdot & & & & \cdot & \cdot & \cdot \\
\cdot & \cdot & \cdot & \cdot & & & & \cdot & \cdot & \cdot \\
0 & 0 & 0 & 0 & \cdot & \cdot & \cdot & a_{k-1} & b_{k-1} & c_{k-1} \\
0 & 0 & 0 & 0 & \cdot & \cdot & \cdot & 0 & a_k & b_k
\end{bmatrix}
\tag{5.A.1}
$$

for $1 \leq k \leq n$. Let D_i denotes the determinant of the principal submatrix consisting of the first i rows and i columns of $\mathbf{B}_k$. Then the determinant D_k of the matrix $\mathbf{B}_k$ in (5.A.1) is obtained by applying the Laplace expansion to the kth (last) row of $\mathbf{B}_k$ and is given by the second order linear recurrence

$$
D_k = b_k D_{k-1} - a_k c_{k-1} D_{k-2},
\tag{5.A.2}
$$

where $D_0 = 1$ and $D_1 = b_1$.

Setting $b_i \equiv 0$ for all $1 \leq i \leq n$, $c_i = \frac{1}{2}(i+1)$ for $1 \leq i \leq (n-1)$, and $a_i = \frac{1}{2}(i-1)$ for $2 \leq i \leq n$ in (5.A.1), it can be verified that $\mathbf{B}_k$ reduces to **A** in (5.5.10). Substituting these values in (5.A.2), the latter becomes

$$
D_k = 0 \cdot D_{k-1} + \frac{k(k-1)}{4} D_{k-2},
\tag{5.A.3}
$$

Table 5.4 Determinant, characteristic polynomial, and eigenvalues of the matrix $\mathbf{A}$ for $2 \leq n \leq 10$

n	Determinant $\|\mathbf{A}\| = \frac{n!}{2^n}$ (n even)	Characteristic polynomial of $\mathbf{A}$	Eigenvalues of $\mathbf{A}$
2	$\frac{1}{2}$	$\lambda^2 + \frac{1}{2}$	$\pm i\,\frac{1}{\sqrt{2}}$
3	0	$\lambda(\lambda^2 + 2)$	$0, \pm\sqrt{2}$
4	$\frac{3}{2}$	$\lambda^4 + 5\lambda^2 + \frac{3}{2}$	$\pm i(2.1632), \pm i(0.5662)$
5	0	$\lambda\left(\lambda^4 + 10\lambda^2 + \frac{23}{2}\right)$	$0, \pm i(2.9452), \pm i(1.1514)$
6	$\frac{45}{4}$	$\lambda^6 + \frac{35}{2}\lambda^4 + 49\lambda^2 + \frac{45}{4}$	$\pm i(3.7517), \pm i(1.7812),$ $\pm i(0.5019)$
7	0	$\lambda\left(\lambda^6 + 28\lambda^4 + 154\lambda^2 + 132\right)$	$0, \pm i(4.5771), \pm i(2.4495),$ $\pm i(1.0297)$
8	$\frac{315}{2}$	$\lambda^8 + 42\lambda^6 + 399\lambda^4 + 818\lambda^2 + \frac{315}{2}$	$\pm i(5.4174), \pm i(3.1486),$ $\pm i(1.5937), \pm i(0.4631)$
9	0	$\lambda\left(\lambda^8 + 60\lambda^6 + 903\lambda^4 + 3590\lambda^2\right)$ $+ \frac{5067}{2}$	$0, \pm i(6.2698), \pm i(3.8730),$ $\pm i(2.1906), \pm i(0.9460)$
10	$\frac{14,175}{4}$	$\lambda^{10} + 165\lambda^8 + 1848\lambda^6 + \frac{25,235}{2}\lambda^4$ $+ \frac{41,877}{2}\lambda^2 + \frac{14,175}{4}$	$\pm i(7.1323), \pm i(4.6165),$ $\pm i(2.8239), \pm i(1.5860),$ $\pm i(0.4363)$

with $D_0 = 1$ and $D_1 = 0$. Iterating (5.A.3) it can be verified that

$$D_k = \begin{cases} \frac{k!}{2^k}, & \text{for even } k \\ 0, & \text{for odd } k. \end{cases} \tag{5.A.4}$$

Thus, $\mathbf{A}$ in (5.5.10) is singular when n is odd. Henceforth we only consider the case when n is even. Refer to Table 5.4 for values of the determinant of $\mathbf{A}$ for $2 \leq n \leq 10$.

The characteristic polynomial of $\mathbf{A}$ is found by setting $b_i = -\lambda_i$, $c_i = \frac{1}{2}(i+1)$, and $a_i = -\frac{1}{2}(i-1)$. In this case, the determinant D_n of B_n in (5.A.1) represents the characteristic polynomial of $\mathbf{A}$ in (5.5.10). Making the above substitutions in (5.A.2), the latter becomes

$$D_k = -\lambda \cdot D_{k-1} + \frac{k(k-1)}{4}D_{k-2}, \tag{5.A.5}$$

with $D_0 = 1$ and $D_1 = -\lambda$. Iterating (5.A.5) leads to the sequence of characteristic polynomial and eigenvalues of $\mathbf{A}$ for $2 \leq n \leq 10$. From this table it clear that the absolute value of the largest eigenvalue increases and the smallest (nonzero) eigenvalue decreases with n. It turns out that the larger eigenvalue corresponds to the high frequency component and the smaller eigenvalues correspond to the low frequency components that make up the solution $\mathbf{q}(t)$ of (5.5.8).

Jordan Canonical Form for **A**

Let $\Lambda \in \mathbb{R}^{n \times n}$ denote the matrix eigenvalues of $\mathbf{A}$ and let $\mathbf{V} \in \mathbb{R}^{n \times n}$, a nonsingular matrix of the corresponding eigenvectors, that is

$$\mathbf{AV} = \mathbf{V}\Lambda \tag{5.A.6}$$

Then

$$\mathbf{A} = \mathbf{V}\Lambda\mathbf{V}^{-1} \tag{5.A.7}$$

and Λ takes a special block diagonal form

$$\mathbf{B}_k = \begin{bmatrix} L_1 & 0 & \cdot & \cdot & \cdot & 0 \\ 0 & L_2 & 0 & \cdot & \cdot & 0 \\ \cdot & 0 & L_3 & 0 & \cdot & 0 \\ \cdot & \cdot & \cdot & \cdot & \cdot & \cdot \\ \cdot & \cdot & \cdot & \cdot & \cdot & \cdot \\ 0 & 0 & 0 & 0 & 0 & L_{\frac{n}{2}} \end{bmatrix} \tag{5.A.8}$$

and

$$\mathbf{L}_i = \begin{bmatrix} 0 & \lambda_i \\ -\lambda_i & 0 \end{bmatrix} \tag{5.A.9}$$

for each complex conjugate pair $\pm i\lambda_i$ of eigenvalues of $\mathbf{A}$, for $1 \le i \le \frac{n}{2}$. The matrix Λ in (5.A.8) is known as the Jordan canonical form of $\mathbf{A}$ (Hirsch et al. 2004).

Solution of (5.5.8)

The general form of the solution $q(t))$ of (5.5.8) is given by

$$\mathbf{q}(t) = e^{\Lambda t}\,\mathbf{q}(0) \tag{5.A.10}$$

Using (5.A.7) in (5.A.10), it can be shown that

$$\mathbf{q}(t) = e^{\mathbf{V}\Lambda\mathbf{V}^{-1}t}\,\mathbf{q}(0)$$
$$= \mathbf{V}\,e^{\Lambda t}\,\mathbf{V}^{-1}\mathbf{q}(0) \tag{5.A.11}$$

or

$$\bar{q}(t) = e^{\Lambda t}\,\bar{q}(0) \tag{5.A.12}$$

where $q(t)$ and $\bar{q}(t)$ are related by the linear transformation

$$\bar{q}(t) = V^{-1}q(0) \tag{5.A.13}$$

By exploring the structure of Λ, it can be verified that

$$e^{\Lambda t} = \begin{bmatrix} e^{L_1 t} & & & \\ & e^{L_2 t} & & \\ & & \ddots & \\ & & & e^{L_{\frac{n}{2}} t} \end{bmatrix}, \tag{5.A.14}$$

where

$$L_i = \begin{bmatrix} c_i & s_i \\ -s_i & c_i \end{bmatrix} \tag{5.A.15}$$

and

$$c_i = \cos(\lambda_i t) \quad \text{and} \quad s_i = \sin(\lambda_i t) \tag{5.A.16}$$

Substituting (5.A.14) through (5.A.16) into (5.A.12), we obtain $\bar{q}(t)$. Clearly, $q(t) = V\bar{q}(t)$ is the solution of (5.A.8). We conclude this appendix with the following

Example 1 (A.1).
Consider the case with $n = 4$ and

$$\frac{dq(t)}{dt} = Aq(t) \tag{5.A.17}$$

with A given by (5.5.8). From Table 5.4, the eigenvalues of A (listed in the increasing order of their absolute values computed using MATLAB) are given by

$$\pm i\lambda_1 = \pm i(0.5662) \quad \text{and} \quad \pm i\lambda_2 = \pm i(2.1662). \tag{5.A.18}$$

From (5.A.14) and (5.A.16), we obtain

$$L_1 = \begin{bmatrix} 0 & 0.5662 \\ -0.5662 & 0 \end{bmatrix} \quad \text{and} \quad L_2 = \begin{bmatrix} 0 & 2.1662 \\ -2.1662 & 0 \end{bmatrix} \tag{5.A.19}$$

and

$$e^{\Lambda t} = \begin{bmatrix} e^{L_1 t} & 0 \\ 0 & e^{L_2 t} \end{bmatrix} \tag{5.A.20}$$

where

$$e^{L_1 t} = \begin{bmatrix} c_1 & s_1 \\ -s_1 & c_1 \end{bmatrix} \quad \text{and} \quad e^{L_2 t} = \begin{bmatrix} c_2 & s_2 \\ -s_2 & c_2 \end{bmatrix} \tag{5.A.21}$$

and

$$c_1 = \cos(0.5662t), \quad c_2 = \cos(2.1662t)$$
$$s_1 = \sin(0.5662t), \quad s_2 = \sin(2.1662t)$$

Hence, $\bar{q}(t) \in \mathbb{R}^4$ is given by

$$\bar{q}_1(t) = c_1\bar{q}_1(0) + s_1\bar{q}_2(0)$$
$$\bar{q}_2(t) = -s_1\bar{q}_1(0) + c_1\bar{q}_2(0)$$
$$\bar{q}_3(t) = c_2\bar{q}_3(0) + s_2\bar{q}_4(0) \tag{5.A.22}$$
$$\bar{q}_4(t) = -s_2\bar{q}_3(0) + c_2\bar{q}_4(0)$$

It can be verified that the matrix of eigenvector $\mathbf{V}$ corresponding to Λ above is given by

$$\mathbf{V} = \begin{bmatrix} -0.8340 & 0 & -0.2413 & 0 \\ 0 & 0.4726 & 0 & 0.5220 \\ -0.0999 & 0 & 0.6723 & 0 \\ 0 & 0.2646 & 0 & -0.4662 \end{bmatrix} \tag{5.A.23}$$

Hence, the solution of (5.A.17) is given by

$$q_1(t) = -0.8340 \cdot \bar{q}_1(t) - 0.2413 \cdot \bar{q}_3(t)$$
$$q_2(t) = 0.4726 \cdot \bar{q}_2(t) + 0.5220 \cdot \bar{q}_4(t)$$
$$q_3(t) = -0.0999 \cdot \bar{q}_1(t) + 0.6723 \cdot \bar{q}_3(t) \tag{5.A.24}$$
$$q_4(t) = 0.2646 \cdot \bar{q}_2(t) - 0.4662 \cdot \bar{q}_4(t)$$

Clearly, the general solution $q_i(t)$ for each i is a linear combination of the harmonic terms $\cos(\lambda_k t)$ and $\sin(\lambda_k t), 1 \leq k \leq \frac{n}{2}$, where the coefficients of the linear combination are given by the elements of the ith row of the matrix $\mathbf{V}$ of eigenvectors of $\mathbf{A}$.

Part II
Applications of Forward Sensitivity Method

Chapter 6
The Gulf of Mexico Problem: Return Flow Analysis

In the late fall and winter, a rhythmic cycle of cold air penetrations into the Gulf of Mexico (GoM) takes place. Whether deep, intermediate or shallow, these penetrations are generally followed by return of modified air to land in response to circulation around an eastward-moving cold anticyclone. Keith Henry, late professor of meteorology at Texas A&M University aptly termed this large-scale process "return flow". Typically, 4–5 of these return-flow events (RFE) occur per month between November and March each year. Field exercise to study the RFEs took place during GUFMEX (Gulf of Mexico Experiment) in 1988 and 1991 (Crisp and Lewis 1992; Lewis 1993; Lewis and Crisp 1992; Liu et al. 1992). The basic components of the RFE are captured in project's logo shown in Fig. 6.1. The schematic depicts a swath of surface air that tracks from the Gulf's northern coastal plain to mid-Gulf where the dark blue color represents the cold dry continental air that is modified as it traverses the relatively warm Gulf waters. The modified air is represented by the light blue color. The dark red swath represents warm, moist tropical air that is sometimes entrained into the stream of modified air. The pink ribbon represents the mixture of modified and tropical air that is returning to the coastal plain. Variations of this theme are myriad—indefinite but large—and this is central to the challenge of accurately forecasting the phenomenon.

A good example of the sea surface temperature (SST) over the Gulf in wintertime is shown in Fig. 6.2. The temperatures are color coded according to the scale in the lower-left corner of the figure. Note that the coldest temperatures (~ 10–$15\,^\circ$C) are found over the water adjoining the coastal plain. The water is quite shallow out to a distance of 100–150 km from shore (labeled shelf water). The very warm SSTs are associated with the Loop current—a clockwise gyre of water that enters the Gulf through the Straits of Yucatan and exits through the Florida Straits. Here the SSTs are $\sim 25\,^\circ$C (the red colored area). As the cold continental air moves southward over the Gulf, with temperatures in the range of 0–10 $^\circ$C, it is warmed and moistened before it turns westward and then northward as it returns to the continent.

© Springer International Publishing Switzerland 2017
S. Lakshmivarahan et al., *Forecast Error Correction using Dynamic Data Assimilation*, Springer Atmospheric Sciences, DOI 10.1007/978-3-319-39997-3_6

Fig. 6.1 Logo for Gulf
of Mexico Experiment

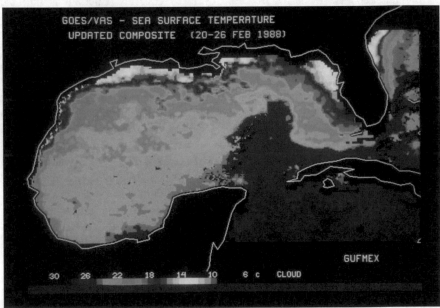

Fig. 6.2 Average SST's over the Gulf of Mexico for the week of February 20–26. Images were collected using the VISSR Atmospheric Sounder (VAS) aboard the geostationary satellite GOES. Color-coded temperature scale (Centigrade) is found in *lower-left corner* of the graphic. (Figure extracted from Lewis et al. 1989)

The difficulty of forecasting water-vapor mixing ratio (grams of water vapor per kilograms of air, the order of 10 g/kg) in return flow over GoM has been well documented in the literature over the past several decades. The following factors have been conjectured to contribute to forecast errors:

1. Absence of routine upper-air observations over the Gulf,
2. Absence of dew point (moisture) measurements on tethered buoys over the Gulf's shelf water,
3. Errors in sea-surface temperature (SST) due to aged data in response to cloud cover, and
4. Inaccuracy in the operational model parameterization of moisture and heat fluxes at the sea-air interface.

Bias, both positive and negative, also has plagued the operational numerical prediction of the mixing ratio, and this aspect of the problem has been especially problematic for the forecaster. The consequence of poor guidance is extreme where forecasts can range from sea fog and stratus cloud when vapor content is low, to shallow cumulus with light showers for intermediate values of the moisture, to cumulonimbus and associated severe weather for large-magnitude vapor mixing ratios.

In an effort to improve forecasts of the RFE, we present a data assimilation strategy based on the FSM. We restrict our investigation to the "outflow phase" of RFE's—that portion of the trajectory over the Gulf of Mexico where the ocean temperature is warmer than the overlying air temperature. In this regime, a convective boundary layer forms—called the mixed layer that can extend to 1–2 km above the sea level (ASL) by the time the initially cold-air moves over the central Gulf (typically 1–2 days after the cold-dry air enters the Gulf). During the "return flow phase", the air begins its return to land and the ocean temperature gradually decreases and a point is reached where the ocean is colder than the overlying air. At this stage, the boundary layer becomes neutral to stable and equations governing the convective boundary layer cease to be valid. But it is well-known among forecasters that the sources of errors in the numerical prediction models are mostly governed by prediction errors during this outflow phase—in essence, when most of the modification of the air mass takes place.

We structure this investigation as follows: (1) observations are presented that define the modification of air as it moves over the Gulf and (2) a comparison of the observations with the idealized mixed-layer process (the convective boundary layer process). The similarity justifies the use of the mixed-layer dynamics in the data assimilation problem discussed in Sect. 6.2. Next we present the equations governing the mixed-layer model and proceed to determine the sensitivity of model output to elements of control. Finally, we perform the FSM using observations from the GUFMEX field study—a case where we have the luxury of having upper-air observations along a Lagrangian trajectory. Most of these observations come from a ship whose speed and direction were the same as outflowing air in Sects. 6.3 and 6.4. We end our investigation with discussion of advantages (and some disadvantages) that come with the use of FSM method.

6.1 Observations in Support of Mixed Layer Modeling

Between February 21 and February 23, 1988, a cold air outflow event occurred over the Gulf of Mexico. A column of the air was observed by taking upper-air observations from a U.S. Coast Guard ship Salvia. It was able to keep pace with the outflowing air for 18 h and observations were taken at 6 h intervals. The track of the ship (at 10–$15\,\mathrm{m\,s^{-1}}$ or 20–30 knots) and the location of observations are shown in the top panel of Fig. 6.3. Note that the trajectory begins to turn westward and then northbound between hours 18 and 36. This is consistent with the typical RFE where the large-scale anticyclone tracks eastward (not shown).

The observations of the potential temperature at points along the trajectory are shown in the bottom panel of Fig. 6.3. The salient features of the profile are gradual warming of the mixed layer (from $\sim$10 to 18 °C) over the 18 h of observations and the increase in height of the mixed layer from $\sim$100 to 1600 m in the 18 h interval (the layer where the potential temperature is constant). Also note that there is a shallow layer near the air-ocean interface (100 m depth) that is labeled the "unstable layer" and a "stable layer" above the mixed layer. In the unstable layer, the potential temperature decreases with height and in the stable layer, the potential temperature increases with height. Thus, there are three rather distinct layers, unstable (0–100 m), mixed layer (100 m to 1 km), and stable layer (above 1–1.5 km). The buoyant elements of air are created in the unstable layer and rise to the "stable layer" (the structure of the air over land that existed before the air moved over the warmer ocean.) The buoyant elements keep breaking down the stable layer from below and the turbulent mixing above this shallow unstable layer gives rise to the mixed layer where the potential temperature is constant and other constituents including water vapor, aerosols like sea salt and pollutants from land, are also uniformly distributed. The evolution of the water vapor (in $\mathrm{g\,kg^{-1}}$) is displayed in Fig. 6.4. Again there is a shallow layer just above the air-sea interface where there is an extremely strong vertical gradient of water vapor (greater at the sea surface). There is a sharp discontinuity in the water vapor above the mixed layer at the same location where the stable layer exists in terms of the potential temperature. Again, the salient feature of the water vapor profile is moistening and deepening.

If we examine the idealized structure of the potential temperature and water vapor associated with the classic mixed layer models of Ball (1960), Lilly (1968), Tennekes and Driedonks (1981), as shown in Fig. 6.6, there is a strong resemblance to the observed thermodynamic structure in our case. This is our justification to use the mixed-layer model dynamics in our study of data assimilation for the February 1988 RFE.

6.2 The Five Variable Model

An air column moves over the ocean, as depicted in Fig. 6.5. The sea surface potential temperature (SST) is known, $\theta_s(t)$. From known SST, we can determine saturation of water vapor mixing ratio at the surface, $q_s(t)$. We assume the warm

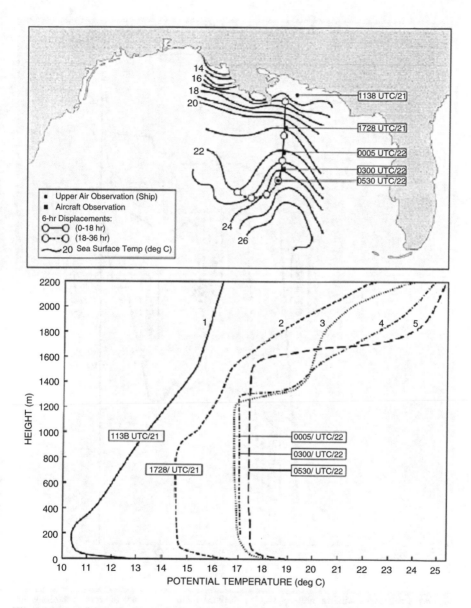

Fig. 6.3 *Panel 1*: The *curved solid line* and its *dashed extension* depict trajectory of surface air that leaves the coastal plain at 1200 UTC (Universal Time), February 21, 1988. The trajectory overlies the sea surface temperature during February 20–21. *Blackened circles* and the *blackened square* show the locations (and times) of upper-air observations taken along this route. *Panel 2*: Profiles of potential temperature (°C) along the trajectory in Panel 1. The numbering of the profiles corresponds to the locations and times of observations shown in Panel 1

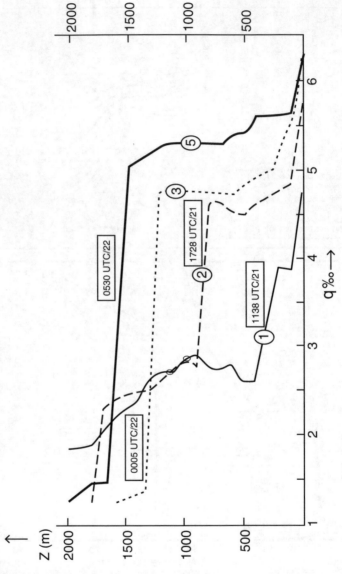

Fig. 6.4 Profiles of mixing ratio $q(\text{g kg}^{-1})$ derived from upper air observations along the trajectory shown in Fig. 6.3. The numbering of the profiles corresponding to the locations and times of observations shown in Fig. 6.3

ocean with colder air above produces a turbulent "mixed" layer—with potential temperature $\theta(t)$ and vapor mixing ratio $q(t)$ constant in this layer. The height $H(t)$ of the air column is increasing with time. Within the column, the temperature increases and vapor increases as turbulent motion transfers heat and water vapor from ocean to atmosphere. The idealized profiles of temperature θ and vapor q are depicted in Fig. 6.6.

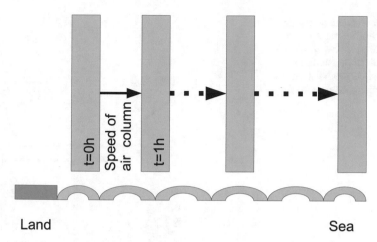

Fig. 6.5 Physical picture

The dynamics of evolution of the five variables $\theta, H, q, \sigma, \mu$ is given by the coupled nonlinear system of ODE's:

$$\frac{d\theta}{dt} = \frac{1+\kappa}{H} C_T V_s (\theta_s - \theta)$$

$$\frac{dH}{dt} = \frac{\kappa}{\sigma} C_T V_s (\theta_s - \theta) + \overline{w}$$

$$\frac{dq}{dt} = \frac{1}{H} \left[C_T V_s (q_s - q) + \frac{\kappa \mu}{\sigma} C_T V_s (\theta_s - \theta) \right] \qquad (6.2.1)$$

$$\frac{d\sigma}{dt} = \gamma_\theta \left(\frac{dH}{dt} - \overline{w} \right) - \frac{d\theta}{dt}$$

$$\frac{d\mu}{dt} = \gamma_q \left(\frac{dH}{dt} - \overline{w} \right) - \frac{dq}{dt}$$

The model parameters are: γ_θ: lapse rate of temperature above mixed layer, γ_q: lapse rate of vapor above mixed layer, κ: entrainment coefficient, $\overline{w}$: large-scale vertical velocity, and $C_T V_s$: exchange coefficient for heat and vapor.

Typical values of the parameter are γ_θ : $3.3 \times 10^{-3} \,°\text{C}\,\text{m}^{-1}$, γ_q : $-1.0 \times 10^{-3} \,\text{g}\,\text{kg}^{-1}\,\text{m}^{-1}$, w : $-1.0\,\text{cm}\,\text{s}^{-1}$, k :$\sim$ $0.2 - 0.3$ [non-dimensional], and $C_T V_s$: $\sim 1.0 \times 10^{-2}\,\text{m}\,\text{s}^{-1}$.

The boundary conditions are $\theta_s(t)$: sea surface temperature and $q_s(t)$: saturated vapor mixing ratio at the sea surface. Boundary conditions are changing along the trajectory. Their values are in the Table 6.1.

We now seek a numerical solution for the five variable model. The variables and parameters follow:

(a) variables $(\theta, H, q, \sigma, \mu)^T = (x_1, x_2, x_3, x_4, x_5)^T = \mathbf{x}$,
(b) parameters $(\kappa, \gamma_\theta, \gamma_q, C_T V_s, \overline{w})^T = (a_1, a_2, a_3, a_4, a_5)^T = \alpha$,

Fig. 6.6 Schematic diagram
of the idealized mixed layer
model displays the profile of
potential temperature on the
right and the profile of water
vapor mixing ratio on the *left*

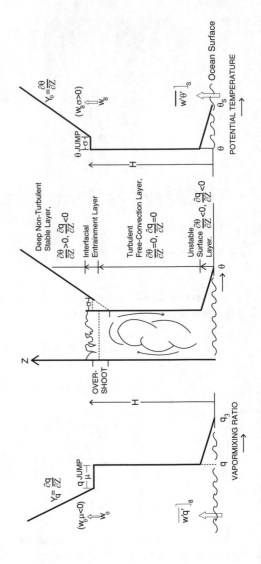

(c) initial   conditions $(x_1(0), x_2(0), x_3(0), x_4(0), x_5(0))^T = (10\,°C, \quad 250\,m,$
$4\,g\,kg^{-1}, 1.5\,°C, -1.5\,g\,kg^{-1})$ and
(d) boundary conditions $(\theta_s(t), q_s(t),)^T = (b_1(t), b_3(t))^T$.

Since the range of values taken by the variables vary widely, it is better to
normalize the variables and parameters.

We normalize model equations using the following constants of normalization:
time t by $T_{norm} = 3600$, $\theta(t)$ by $\theta_{norm} = 24$, $H(t)$ by $H_{norm} = 1900$, $q(t)$ by
$q_{norm} = 5$, $\sigma(t)$ by $\sigma_{norm} = 1.5$, and $\mu(t)$ by $\mu_{norm} = -4$.

After normalization, we change the variables and simplify model and sensitivity
equations, by introducing the following constants:

Table 6.1 Measurements

Time (h)	θ_s (C)	q_s (g/kg)
0	16.50	11.40
1	17.00	11.77
2	18.00	12.54
3	19.30	13.60
4	20.10	14.29
5	20.60	14.74
6	21.00	15.11
7	21.20	15.29
8	21.25	15.34
9	21.20	15.29
10	21.50	15.58
11	21.60	15.67
12	21.70	15.77
13	21.90	15.96
14	22.00	16.06
15	22.50	16.55
16	23.00	17.06
17	24.00	18.12
18	24.50	18.67

$$A = \frac{T_{norm}}{H_{norm}}, \quad B = \frac{\theta_{norm}}{\sigma_{norm}}, \quad X = \frac{H_{norm}}{\sigma_{norm}},$$

$$H = H_{norm}, \quad M = \mu_{norm}, \quad Y = \frac{q_{norm}}{\mu_{norm}}.$$

(6.2.2)

The resulting set of five equations after normalization are given by

$$\dot{x}_1 = f_1(k,\alpha) = \frac{A\,a_4\,(b_1 - x_1(k))\,(a_1 + 1)}{x_2(k)}$$

$$\dot{x}_2 = f_2(k,\alpha) = A\left[a_5 + \frac{B\,a_1\,a_4\,(b_1 - x_1(k))}{x_4(k)}\right]$$

$$\dot{x}_3 = f_3(k,\alpha) = \frac{A}{x_2(k)}\left[a_4\,(b_3 - x_3(k)) + \frac{B\,a_1\,a_4\,f_5(k)\,(b_1 - x_1(k))}{Y\,x_4(k)}\right]$$

$$\dot{x}_4 = f_4(k,\alpha) = -A\,a_4\left[\frac{a_1 + 1}{x_2(k)} - \frac{X\,a_1\,a_2}{x_4(k)}\right](b_1 - x_1(k))$$

$$\dot{x}_5 = f_5(k,\alpha) = -A\,a_4\left[\frac{Y\,(b_3 - x_3(k))}{x_2(k)} + \frac{B\,H\,a_1}{x_4(k)}\left(x_5(k) - \frac{a_3}{M}\right)(b_1 - x_1(k))\right]$$

(6.2.3)

A plot of the typical solution of (6.2.3) are given in Figs. 6.7, 6.8, 6.9, 6.10, and 6.11.

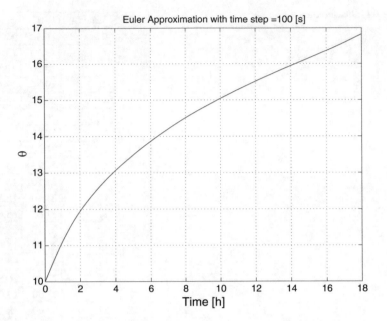

Fig. 6.7 Potential temperature θ as x_1

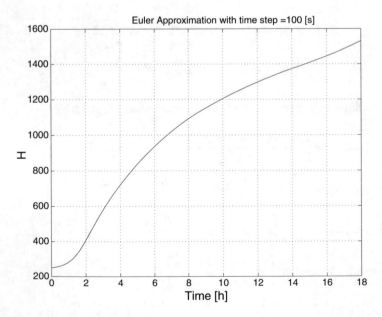

Fig. 6.8 Mixed layer height H denoted as x_2

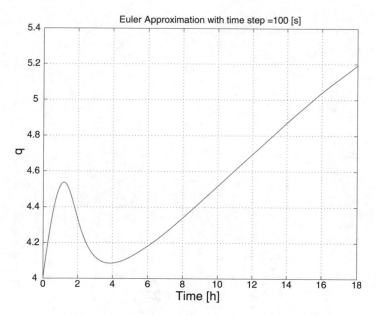

Fig. 6.9 Water vapor concentration in mixed layer q as x_3

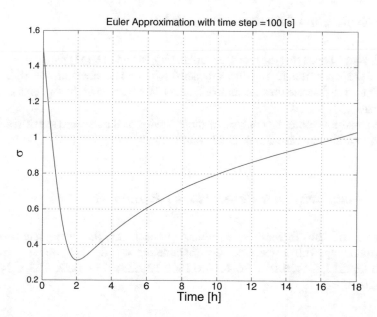

Fig. 6.10 Jump in potential temperature at the top of the mixed layer σ as x_4

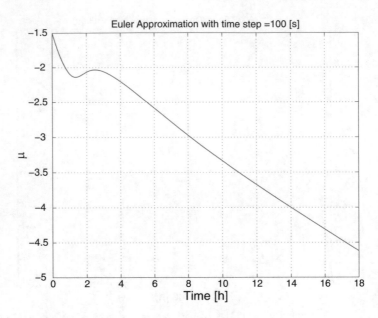

Fig. 6.11 Jump in water vapor content at the top of the mixed layer μ as x_5

6.3   Evolution of Forward Sensitivities

Recall from Chap. 4 that the model Jacobians $\mathbf{D_x(M)}$ and $\mathbf{D_\alpha(M)}$ play a fundamental role in the evolution of the forward sensitivities. The elements of $\mathbf{D_x(M)}$ and $\mathbf{D_\alpha(M)}$ for the discrete time normalized model $\mathbf{M}$ are given in the Appendix "Model Jacobians".

We now move onto the analysis of the behavior of the forward sensitivities with respect to both parameters and initial conditions.

6.3.1   Sensitivity with Respect to the Parameters

Evolution of the forward sensitivities of each of the five state variables x_1 through x_5 with respect to five parameters α_1 through α_5 are given in Figs. 6.12, 6.13, 6.14, 6.15, 6.16, 6.17, 6.18, 6.19, 6.20, 6.21, 6.22, 6.23, 6.24, 6.25, and 6.26.

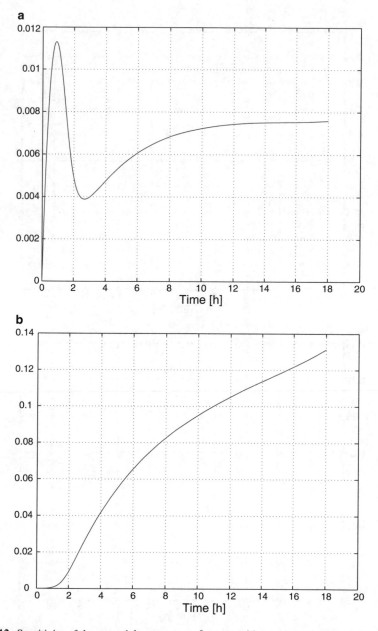

Fig. 6.12 Sensitivity of the potential temperature θ, as x_1, with respect to the parameters α_1 and α_2. (**a**) Sensitivity x_1 w.r.t. α_1. (**b**) Sensitivity of x_1 w.r.t. α_2

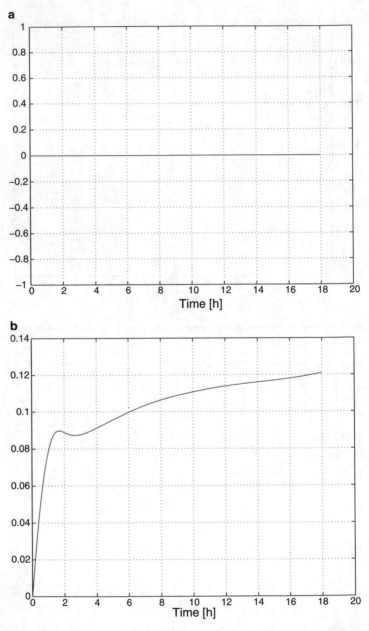

Fig. 6.13 Sensitivity of the potential temperature θ, as x_1, with respect to the parameters α_3 and α_4. (**a**) Sensitivity of x_1 w.r.t. α_3. (**b**) Sensitivity of x_1 w.r.t. α_4

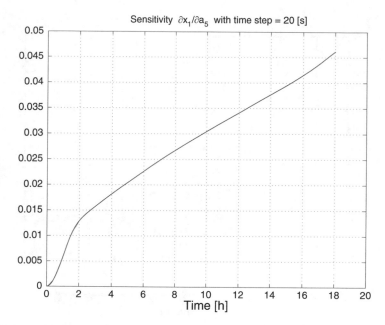

Fig. 6.14 Sensitivity of the potential temperature θ, as x_1, with respect to the parameter α_5

6.3.2 Sensitivity with Respect to Initial Conditions

(1) Sensitivity of the potential temperature θ, as x_1, with respect to the initial conditions: Results are given in Figs. 6.27, 6.28, and 6.29. We can see that x_1 is the most sensitive to its initial conditions just at the beginning of the simulation, and that this sensitivity decrease in time. It is sensitive to $x_2(0)$ and $x_4(0)$, but its sensitivity to them grows in time from zero, only to reach its maximum during the first 4 h of simulation.

(2) Sensitivity of height H, as x_2, with respect to the initial conditions: Results are given in Figs. 6.30, 6.31, and 6.32. We can see that x_2 is the most sensitive to its initial conditions just at the beginning of the simulation, and that this sensitivity decrease in time, to reach zero at about 4 h. It is sensitive to $x_1(0)$ and $x_4(0)$, but its sensitivity to them grows in time from zero, only to reach its maximum during the first 4 h of simulation, and then decrease at about sixth hours of simulation to stay very close to zero.

(3) Plots of sensitivity of the water vapor concentration q, as x_3, with respect to the initial conditions are in Figs. 6.33, 6.34, and 6.35. We can see that x_3 is the most sensitive to its initial conditions just at the beginning of the simulation, and that this sensitivity decrease in time, to reach zero at the end f the simulation. We can also notice that it its sensitivity to remaining initial conditions reaches its maximum during the first 2–4 h of the simulation.

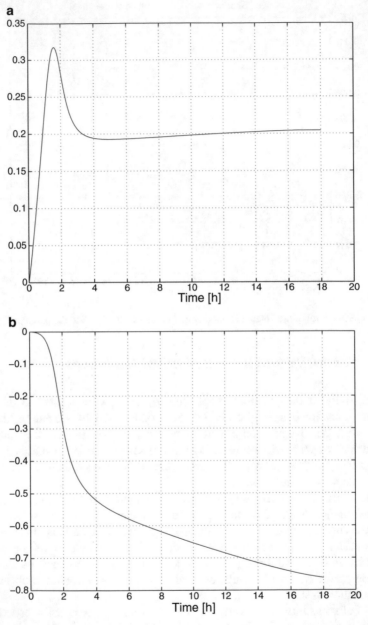

Fig. 6.15 Sensitivity of the height H, as x_2, with respect to the parameters α_1 and α_2. (**a**) Sensitivity x_2 w.r.t. α_1. (**b**) Sensitivity of x_2 w.r.t. α_2

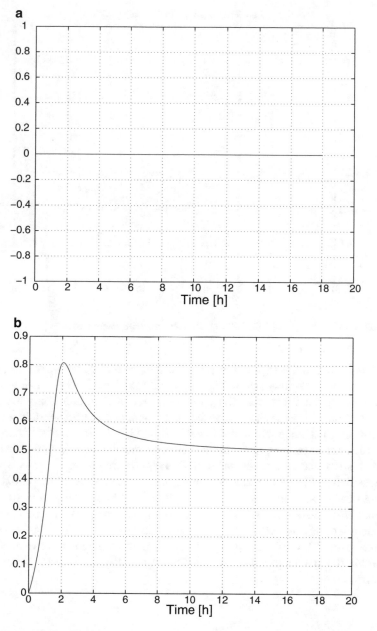

Fig. 6.16 Sensitivity of the height H, as x_2, with respect to the parameters α_3 and α_4. (**a**) Sensitivity of x_2 w.r.t. α_3. (**b**) Sensitivity of x_2 w.r.t. α_4

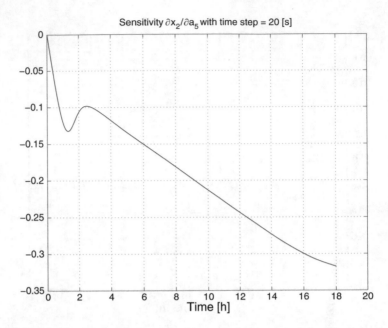

Fig. 6.17 Sensitivity of the height H, as x_2, with respect to the parameter α_5

(4) Plots of sensitivity of jump in potential temperature σ, as x_4, with respect to the initial conditions are in Figs. 6.36, 6.37, and 6.38. The sensitivity of x_4 starts from one for its sensitivity to its own initial conditions, only to reach its maximum within the first 2 h of the simulation, and to became only a fraction of its original value within the first 4 h of simulation. Sensitivities to $x_1(0)$ and $x_2(0)$ start from zero, reach their respective maximums within the first 4 h of simulation, and drop to almost zero for the remaining time.

(5) Sensitivity of jump in water content μ, as x_5, with respect to the initial conditions can be seen in Figs. 6.39, 6.40, and 6.41. The sensitivity of x_5 starts from one for its sensitivity to its own initial conditions, and gradually became only a fraction of its original value during the duration of the simulation. Sensitivities to $x_1(0), x_2(0), x_3(0)$ and $x_4(0)$ start from zero, reach their respective maximums within the first 4 h of simulation, and then gradually decrease in the remaining time.

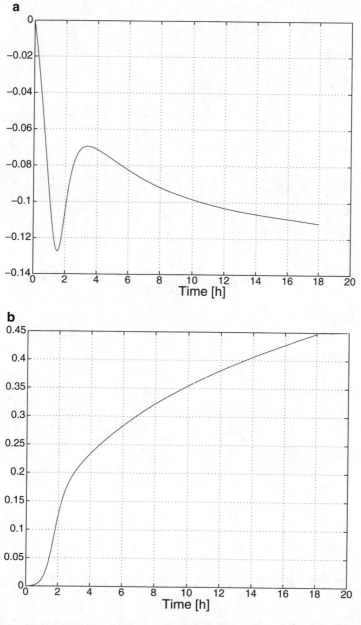

Fig. 6.18 Sensitivity of the water vapor concentration q, as x_3, with respect to the parameters α_1 and α_2. (**a**) Sensitivity x_3 w.r.t. α_1. (**b**) Sensitivity of x_3 w.r.t. α_2

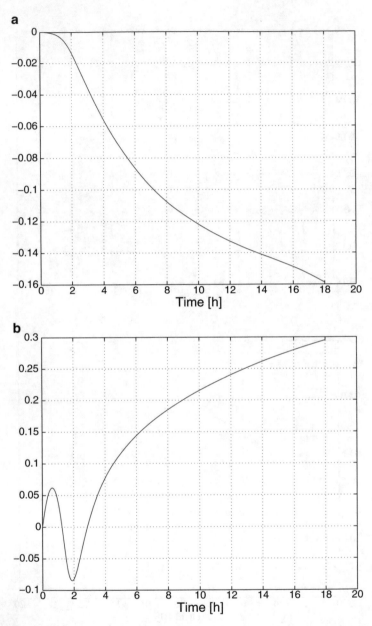

Fig. 6.19 Sensitivity of the water vapor concentration q, as x_3, with respect to the parameters α_3 and α_4. (**a**) Sensitivity of x_3 w.r.t. α_3. (**b**) Sensitivity of x_3 w.r.t. α_4

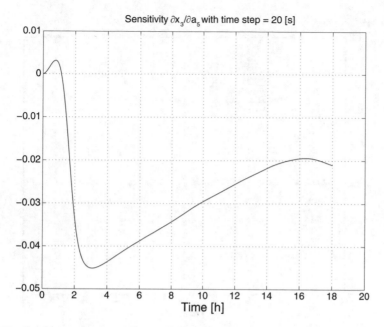

Fig. 6.20 Sensitivity of the water vapor concentration q, as x_3, with respect to the parameter α_5

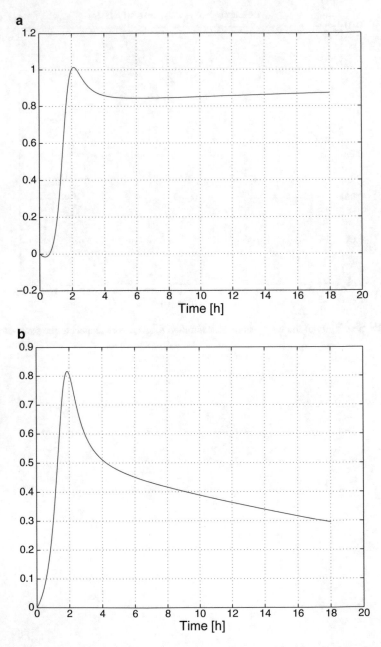

Fig. 6.21 Sensitivity of the potential temperature jump σ, as x_4, with respect to the parameters α_1 and α_2. (**a**) Sensitivity x_4 w.r.t. α_1. (**b**) Sensitivity of x_4 w.r.t. α_2

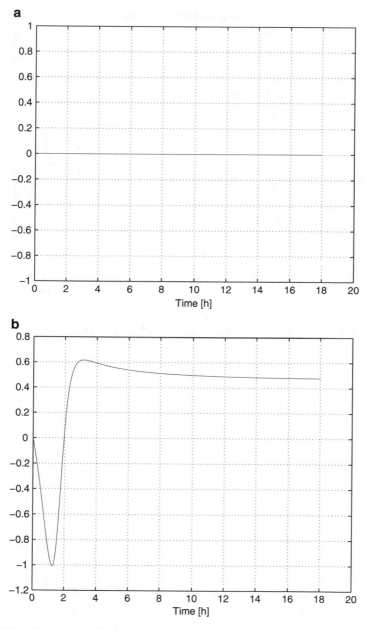

Fig. 6.22 Sensitivity of the potential temperature jump σ, as x_4, with respect to the parameters α_3 and α_4. (**a**) Sensitivity of x_4 w.r.t. α_3. (**b**) Sensitivity of x_4 w.r.t. α_4

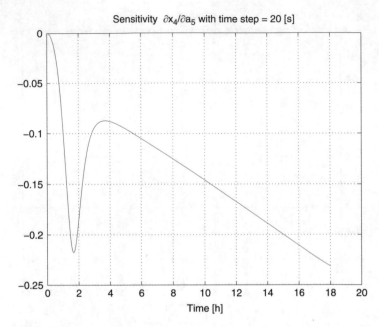

Fig. 6.23 Sensitivity of the potential temperature jump σ, as x_4, with respect to the parameter α_5

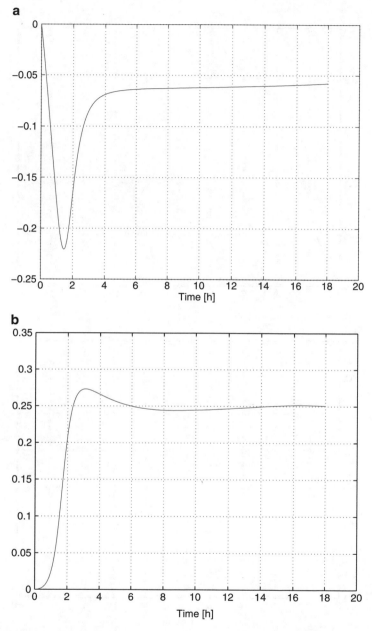

Fig. 6.24 Sensitivity of the water vapor content μ, as x_5, with respect to the parameters α_1 and α_2. (**a**) Sensitivity x_5 w.r.t. α_1. (**b**) Sensitivity of x_5 w.r.t. α_2

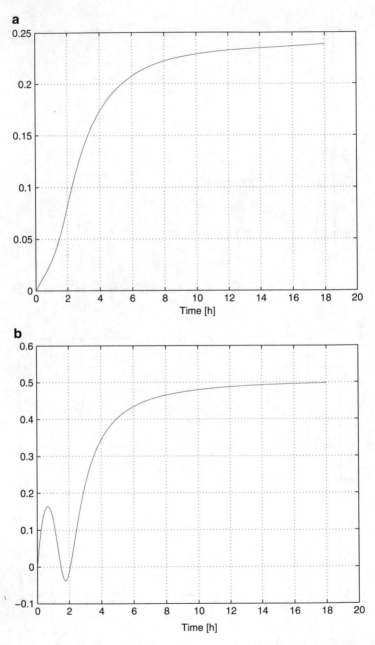

Fig. 6.25 Sensitivity of the water vapor content μ, as x_5, with respect to the parameters α_3 and α_4. (**a**) Sensitivity of x_5 w.r.t. α_3. (**b**) Sensitivity of x_5 w.r.t. α_4

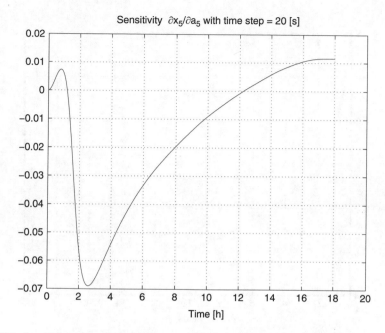

Fig. 6.26 Sensitivity of the water vapor content μ, as x_5, with respect to the parameter α_5

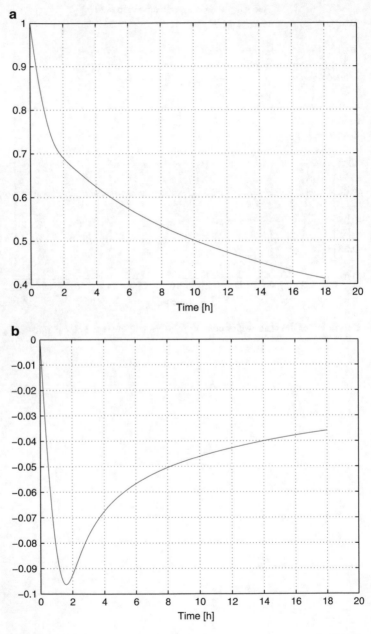

Fig. 6.27 Sensitivity of the potential temperature θ, as x_1, with respect to the initial conditions $x_1(0)$ and $x_2(0)$. (**a**) Sensitivity of x_1 w.r.t. $x_1(0)$. (**b**) Sensitivity of x_1 w.r.t. $x_2(0)$

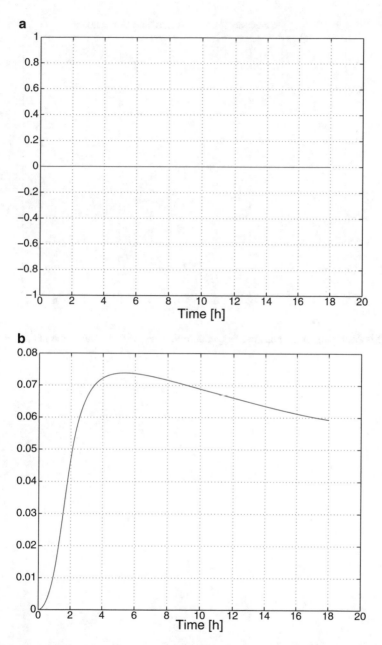

Fig. 6.28 Sensitivity of the potential temperature θ, as x_1, with respect to the initial conditions $x_3(0)$ and $x_4(0)$. (**a**) Sensitivity of x_1 w.r.t. $x_3(0)$. (**b**) Sensitivity of x_1 w.r.t. $x_4(0)$

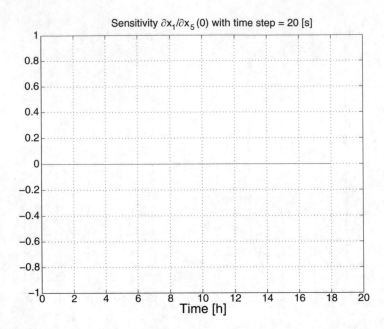

Fig. 6.29 Sensitivity of the potential temperature θ, as x_1, with respect to the initial conditions $x_5(0)$

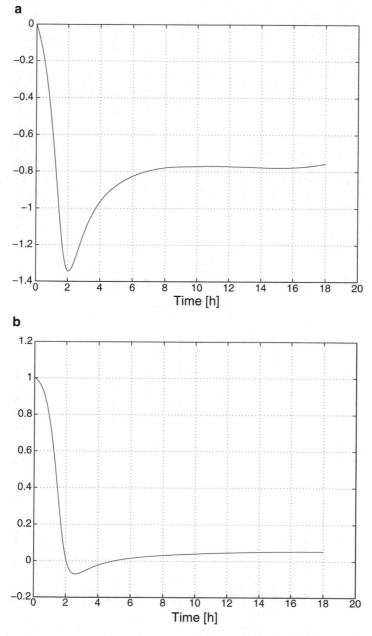

Fig. 6.30 Sensitivity of the height H, as x_2, with respect to the initial conditions $x_1(0)$ and $x_2(0)$.
(**a**) Sensitivity of x_2 w.r.t. $x_1(0)$. (**b**) Sensitivity of x_2 w.r.t. $x_2(0)$

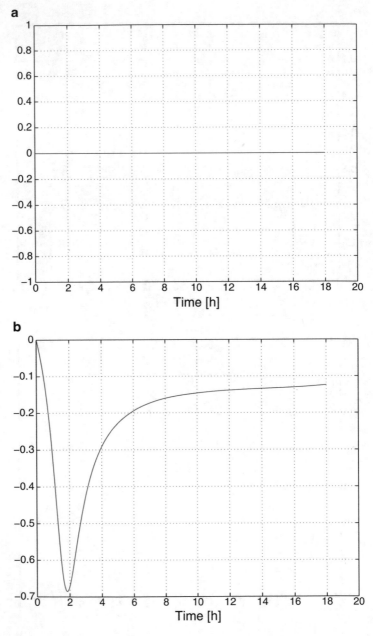

Fig. 6.31 Sensitivity of the height H, as x_2, with respect to the initial conditions $x_3(0)$ and $x_4(0)$. (**a**) Sensitivity of x_2 w.r.t. $x_3(0)$. (**b**) Sensitivity of x_2 w.r.t. $x_4(0)$

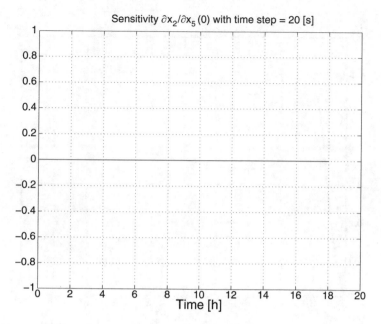

Fig. 6.32 Sensitivity of the height H, as x_2, with respect to the initial conditions $x_5(0)$

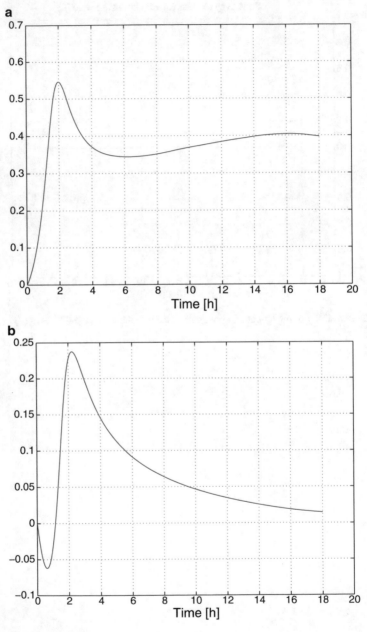

Fig. 6.33 Sensitivity of the water vapor concentration q, as x_3, with respect to the initial conditions $x_1(0)$ and $x_2(0)$. (**a**) Sensitivity of x_3 w.r.t. $x_1(0)$. (**b**) Sensitivity of x_3 w.r.t. $x_2(0)$

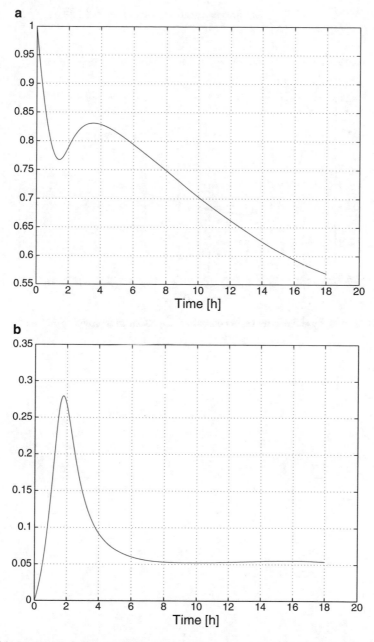

Fig. 6.34 Sensitivity of the water vapor concentration q, as x_3, with respect to the initial conditions $x_3(0)$ and $x_4(0)$. (a) Sensitivity of x_3 w.r.t. $x_3(0)$. (b) Sensitivity of x_3 w.r.t. $x_4(0)$

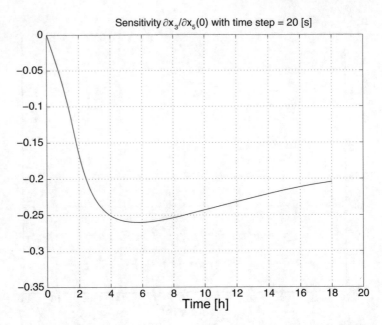

Fig. 6.35 Sensitivity of the water vapor concentration q, as x_3, with respect to the initial conditions $x_5(0)$

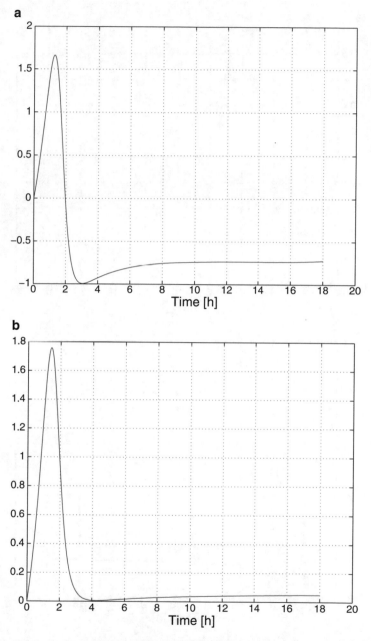

Fig. 6.36 Sensitivity of the jump in potential temperature σ, as x_4, with respect to the initial conditions $x_1(0)$ and $x_2(0)$. (**a**) Sensitivity of x_4 w.r.t. $x_1(0)$. (**b**) Sensitivity of x_4 w.r.t. $x_2(0)$

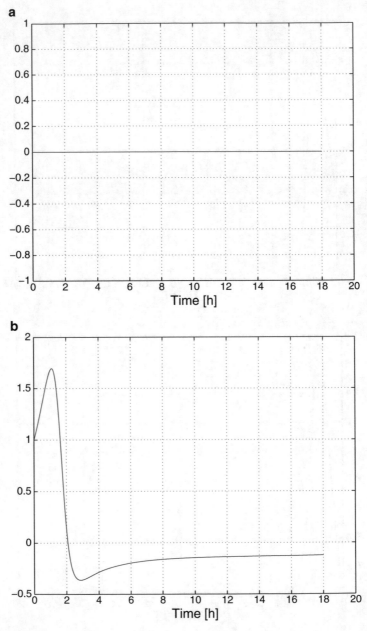

Fig. 6.37 Sensitivity of the jump in potential temperature σ, as x_4, with respect to the initial conditions $x_3(0)$ and $x_4(0)$. (**a**) Sensitivity of x_4 w.r.t. $x_3(0)$. (**b**) Sensitivity of x_4 w.r.t. $x_4(0)$

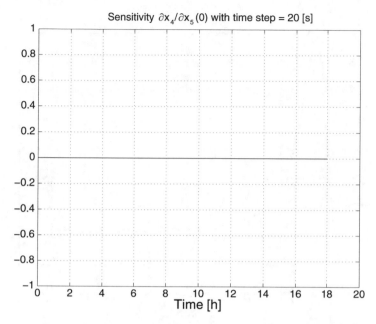

Fig. 6.38 Sensitivity of the jump in potential temperature σ, as x_4, with respect to the initial conditions $x_5(0)$

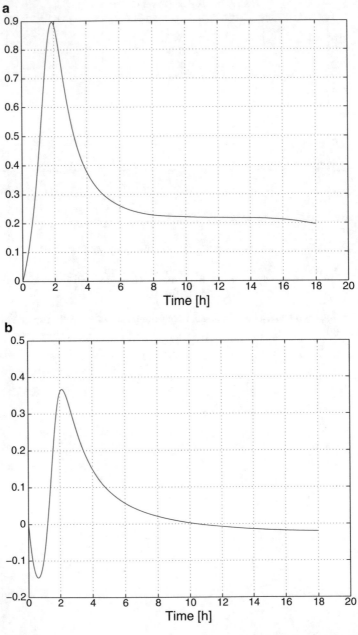

Fig. 6.39 Sensitivity of the jump in water content μ, as x_5, with respect to the initial conditions $x_1(0)$ and $x_2(0)$. (**a**) Sensitivity of x_5 w.r.t. $x_1(0)$. (**b**) Sensitivity of x_5 w.r.t. $x_2(0)$

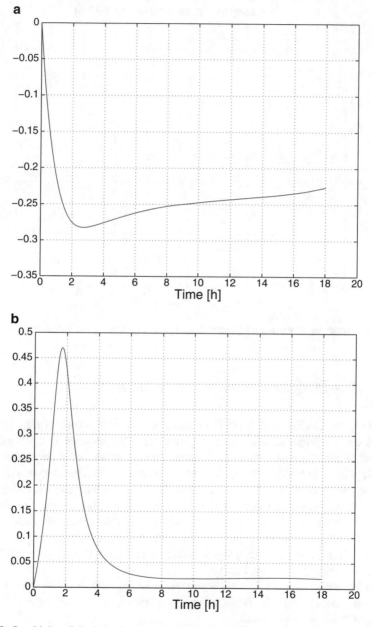

Fig. 6.40 Sensitivity of the jump in water content μ, as x_5, with respect to the initial conditions $x_3(0)$ and $x_4(0)$. (**a**) Sensitivity of x_5 w.r.t. $x_3(0)$. (**b**) Sensitivity of x_5 w.r.t. $x_4(0)$

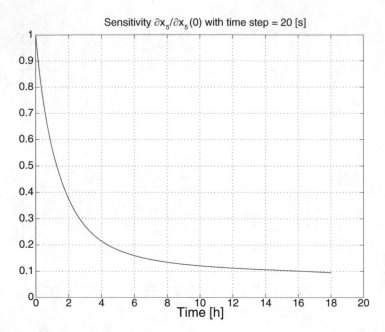

Fig. 6.41 Sensitivity of the jump in water content μ, as x_5, with respect to the initial conditions $x_5(0)$

6.3.3   Sensitivity at the End of the Simulation

Data shown in the Tables 6.2 and 6.3 illustrates how final results of the simulation depend on the initial conditions and parameters. We can see that the potential temperature after 18 h of simulation depends strongly on its initial conditions, initial conditions of σ and parameter γ_θ. The height of the mixed layer depends almost as much on parameter κ as on initial value of the potential temperature, but does not depend on the moisture values at the beginning of the simulation. The deviation of moisture content $x_3 = q$ depends strongly on its initial conditions, and value of $\Delta x_4 = \Delta c_T\, V_s$, but does not depend on the initial height of the mixed layer, or initial value of σ. The jump μ in water vapor content at the top of the mixed layer depends strongly on the parameters κ, and γ_θ.

   This information could be used for pinpointing improvements to the final value of a given model variable by indicating which parameters and initial conditions play a major role in its development.

Table 6.2 Changes in forecast in response to uncertainty in the model parameters—elements of the control vector shown after 18 h of simulation [$(\alpha_1 = \kappa, \alpha_2 = \gamma_\theta, \alpha_3 = \gamma_q, \alpha_4 = c_T V_s, \alpha_5 = \overline{w})$ and $(\theta = x_1, H = x_2, q = x_3, \sigma = x_4, \mu = x_5)$]

Δf	$\Delta\alpha_1$	$\Delta\alpha_2$	$\Delta\alpha_3$	$\Delta\alpha_4$	$\Delta\alpha_5$
Δx_1	0.0425	0.1338	0	0.0407	−0.0388
Δx_2	103.9915	−70.5913	0	15.3360	24.2610
Δx_3	−0.0470	0.0334	0.0009	0.1059	−0.0102
Δx_4	0.3006	0.0185	0	0.0099	0.0120
Δx_5	0.1722	−0.1632	0.0005	0.0238	−0.0300

Table 6.3 Changes in forecast in response to uncertainty in the initial conditions—elements of the control vector shown after 18 h of simulation ($\theta = x_1, H = x_2, q = x_3, \sigma = x_4, \mu = x_5$)

Δf	$\Delta x_1(0)$	$\Delta x_2(0)$	$\Delta x_3(0)$	$\Delta x_4(0)$	$\Delta x_5(0)$
Δx_1	0.6933	−0.1209	0	0.3324	0
Δx_2	−115.6176	16.4567	0	−63.5384	0
Δx_3	0.0459	−0.0511	0.2231	0.0062	0.0009
Δx_4	−0.0748	0.0102	0	−0.0421	0
Δx_5	−0.0831	0.0271	0.0001	−0.0119	0.0000

6.4 Data Assimilation Using Forward Sensitivity Method: Numerical Experiments

We conducted several numerical experiments to improve control vector. In order to do that, we have generated "observations" by running the model with the control vector listed above in Sect. 6.1.

Then, we changed the values of the elements of the control vector and ran the model again in a spirit of the twin experiment to retrieve original control using FSM. The procedure can be described as follows:

(1) use original control vector from Lewis (2007)
(2) run model (6.2.3) forward in time to generate "observations"
(3) modify control vector
(4) run model (6.2.3) forward in time to generate the forecast with the modified control vector
(5) calculate forward sensitivities of $\mathbf{x}(t)$ with respect to $\mathbf{x}(0)$ and $\boldsymbol{\alpha}$
(6) calculate forecast error $\mathbf{e}_F$
(7) calculate correction to the control $\delta\mathbf{c}$
(8) apply correction to control vector
(9) repeat steps (4)–(8), if control vector differs from the original control vector by more than a set norm

In the following we provide a set of three examples, many more are contained in the report by Jabrzemski and Lakshmivarahan (2011).

6.4.1   Observations at Multiple Time

(1) The goal of this experiment is to retrieve the initial conditions using three observations at about 1 min, 500 min (8.7 h) and 1000 min (16.7 h) from the beginning of the simulation which correspond to 2nd, 1500nd and 3000th steps of the model iteration and results can be seen in Figs. 6.42, 6.43 and 6.44. The forecast trajectory from the perturbed initial conditions is given by dotted lines and the corrected forecast using FSM is given by the unbroken lines. The diamond symbols represent the observations. The match between the forecast and the observation is very good as is also evidenced by the reduction in the cost function J.
(2) Retrieval of initial conditions and parameters with observations at about 83 min, 166 min (2.7 h) and 415 min (6.9 h), with time step 100 s (which correspond to model iteration: 50, 100 and 249) The results are given in Figs. 6.45, 6.46 and 6.47.
(3) Retrieval of initial conditions and parameters with observations at 1 min, 50 min, 8.3 h, and 17.2 h corresponding to model iteration: 2, 150, 1500, 2000 and 3100. The results are given in Figs. 6.48, 6.49 and 6.50.

6.5   Problems

Problem 6.1. Run the model (6.2.3) using the values of control—initial condition and parameters and generate observation vector $z = x + v$, where $v \sim (N(0, \sigma))$ with σ being a diagonal matrix given by

$$\sigma = \text{Diag}(\sigma_1^2, \sigma_2^2, \sigma_3^2, \sigma_4^2, \sigma_5^2). \tag{6.5.1}$$

Recall there are five initial conditions and five parameters.

1. Conduct a series of FSM based experiments by starting from incorrect control and varying the number and distribution of observations and produce plots similar to these in Figs. 6.42, 6.49, and 6.44.
2. Compute the eigenvalues and condition number of $(\mathbf{H}^T \mathbf{H})$ or $(\mathbf{H}\mathbf{H}^T)$.

Note: When these matrices are ill-conditioned, consider using Tikhonov's regularization method to get around this difficulty (refer to Lewis et al. 2006, Chap. 5).

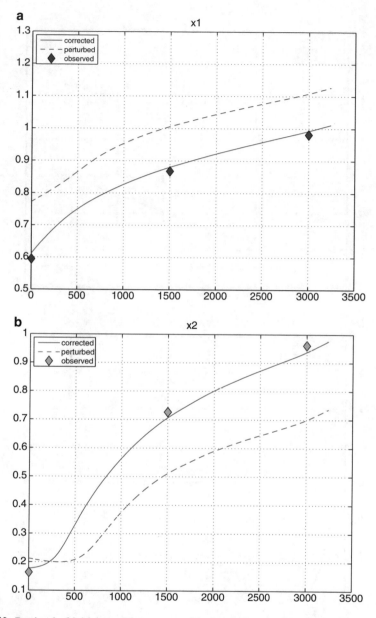

Fig. 6.42 Retrieval of initial conditions x_1, x_2 with observations at about 1 min, 500 min (8.7 h) and 1000 min (16.7 h). (**a**) x_1. (**b**) x_2

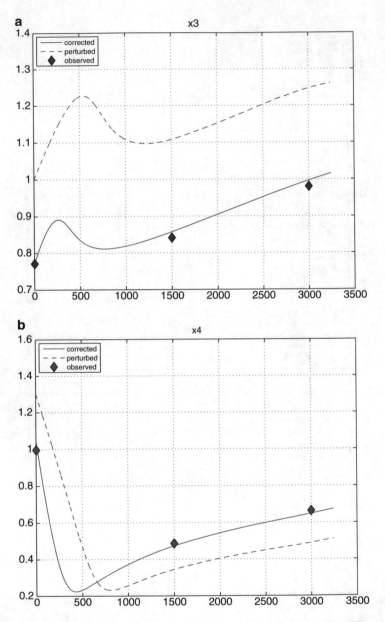

Fig. 6.43 Retrieval of initial conditions x_3, x_4 with observations at about 1 min, 500 min (8.7 h) and 1000 min (16.7 h). (**a**) x_3. (**b**) x_4

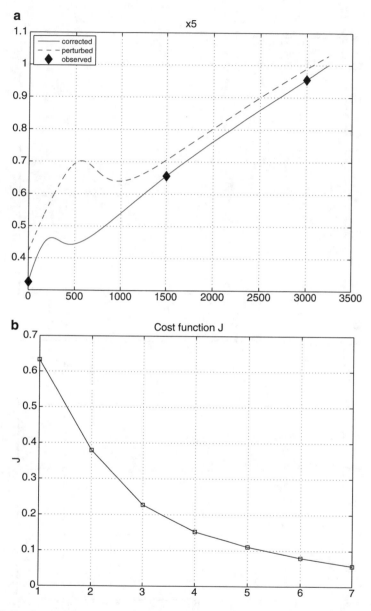

Fig. 6.44 Retrieval of initial conditions x_5 with observations at about 1 min, 500 min (8.7 h) and 1000 min (16.7 h). Cost function J for consecutive iterations during the retrieval. (**a**) x_5. (**b**) J

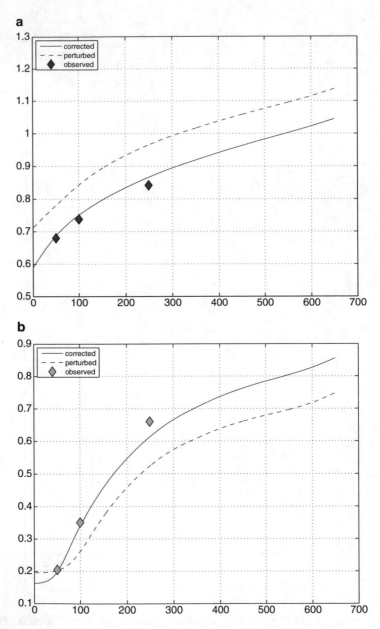

Fig. 6.45 Retrieval of initial conditions x_1, x_2 and parameters observations at about 83 min, 166 min (2.7 h) and 415 min (6.9 h), with time step 100 s. (**a**) x_1. (**b**) x_2

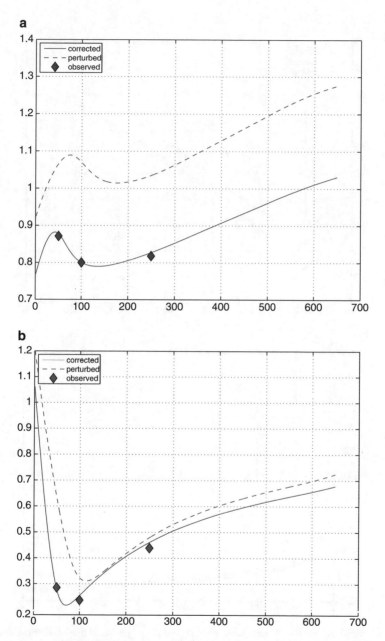

Fig. 6.46 Retrieval of initial conditions x_3, x_4 with observations at about 83 min, 166 min (2.7 h) and 415 min (6.9 h), with time step 100 s. (**a**) x_3. (**b**) x_4

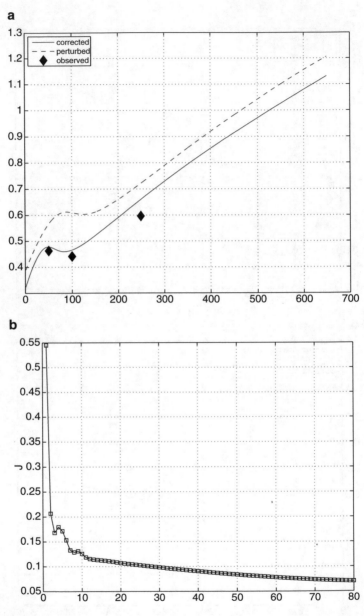

Fig. 6.47 Retrieval of initial conditions x_5 observations at about 83 min, 166 min (2.7 h) and 415 min (6.9 h), with time step 100 s. Cost function J for consecutive iterations during the retrieval. (**a**) x_5. (**b**) J

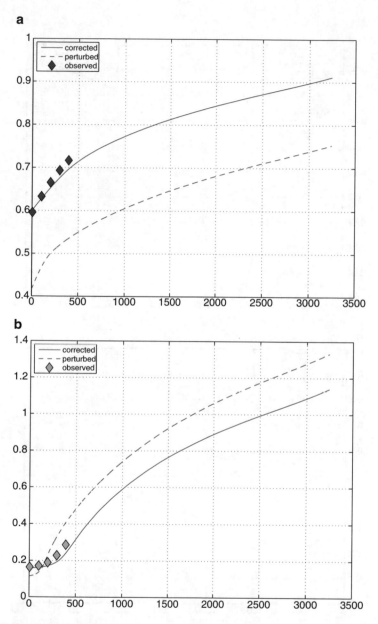

Fig. 6.48 Retrieval of initial conditions x_1, x_2 and parameters with observations at about 1, 34, 66, 98 and 131 min, with time step 100 s. (**a**) x_1. (**b**) x_2

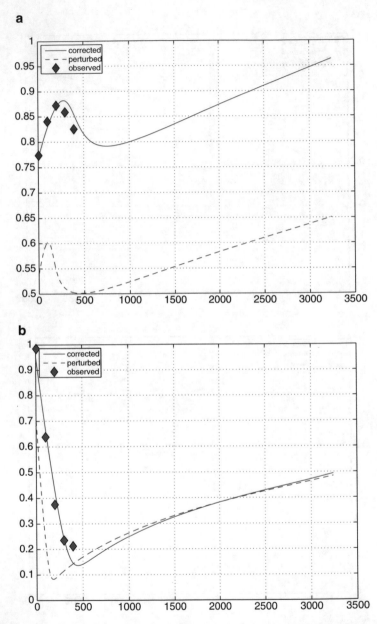

Fig. 6.49 Retrieval of initial conditions x_3, x_4 with observations at about 1, 34, 66, 98 and 131 min. (**a**) x_3. (**b**) x_4

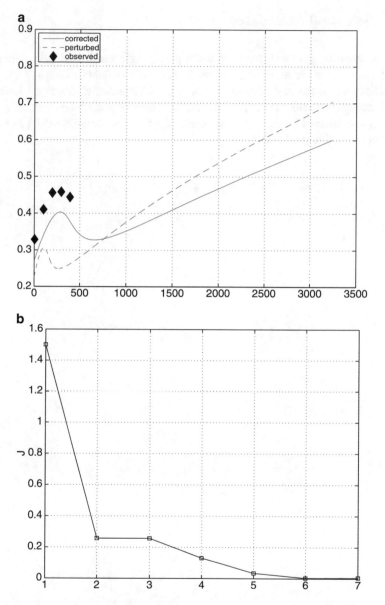

Fig. 6.50 Retrieval of initial conditions x_5 with observations at about 1, 34, 66, 98 and 131 min. Cost function J for consecutive iterations during the retrieval. (**a**) x_5. (**b**) J

6.6   Notes and References

This chapter follows closely the technical report by Jabrzemski and Lakshmivarahan (2011). The air-sea interaction problem described in details in Chaps. 2 and 3 are closely related to the first equation in the five variable model described in (6.2.1). We encourage the reader to vary the number and distribution of the observations and perform the forecast correction experiments for the five variable model much like experiments in Chaps. 2 and 3.

Appendix

Model Jacobians

Elements of the Jacobian $D_x(M)$

$$D_x(1, 1) = -\frac{A\, a_4\, (a_1 + 1)}{x_2(t)}$$

$$D_x(1, 2) = -\frac{A\, a_4\, (b_1 - x_1(t))\, (a_1 + 1)}{x_2(t)^2}$$

$$D_x(1, 3) = 0$$

$$D_x(1, 4) = 0$$

$$D_x(1, 5) = 0$$

$$D_x(2, 1) = -\frac{A\, B\, a_1\, a_4}{x_4(t)}$$

$$D_x(2, 2) = 0$$

$$D_x(2, 3) = 0$$

$$D_x(2, 4) = -\frac{A\, B\, a_1\, a_4\, (b_1 - x_1(t))}{x_4(t)^2}$$

$$D_x(2, 5) = 0$$

$$D_x(3, 1) = -\frac{A\, B\, a_1\, a_4\, x_5(t)}{Y\, x_2(t)\, x_4(t)}$$

$$D_x(3, 2) = -\frac{A}{x_2(t)^2}\left[a_4\, (b_3 - x_3(t)) + \frac{B\, a_1\, a_4\, x_5(t)\, (b_1 - x_1(t))}{Y\, x_4(t)} \right]$$

$$D_x(3, 3) = -\frac{A\, a_4}{x_2(t)}$$

$$D_x(3, 4) = -\frac{A\, B\, a_1\, a_4\, x_5(t)\, (b_1 - x_1(t))}{Y\, x_2(t)\, x_4(t)^2}$$

$$D_x(3, 5) = \frac{A\, B\, a_1\, a_4\, (b_1 - x_1(t))}{Y\, x_2(t)\, x_4(t)}$$

$$\mathbf{D_x}(4, 1) = A\, a_4 \left[\frac{a_1 + 1}{x_2(t)} - \frac{X\, a_1\, a_2}{x_4(t)} \right]$$

$$\mathbf{D_x}(4, 2) = \frac{A\, a_4\, (b_1 - x_1(t))\, (a_1 + 1)}{x_2(t)^2}$$

$$\mathbf{D_x}(4, 3) = 0$$

$$\mathbf{D_x}(4, 4) = -\frac{A\, X\, a_1\, a_2\, a_4\, (b_1 - x_1(t))}{x_4(t)^2}$$

$$\mathbf{D_x}(4, 5) = 0$$

$$\mathbf{D_x}(5, 1) = \frac{A\, B\, a_1\, a_4}{x_4(t)} \left[\frac{x_5(t)}{x_2(t)} - \frac{H\, a_3}{M} \right]$$

$$\mathbf{D_x}(5, 2) = \frac{A\, a_4}{x_2(t)^2} \left[Y\, (b_3 - x_3(t)) + \frac{B\, a_1\, x_5(t)\, (b_1 - x_1(t))}{x_4(t)} \right]$$

$$\mathbf{D_x}(5, 3) = \frac{A\, Y\, a_4}{x_2(t)}$$

$$\mathbf{D_x}(5, 4) = \frac{A\, B\, a_1\, a_4}{x_4(t)^2} \left[\frac{x_5(t)}{x_2(t)} - \frac{H\, a_3}{M} \right] (b_1 - x_1(t))$$

$$\mathbf{D_x}(5, 5) = -\frac{A\, B\, a_1\, a_4\, (b_1 - x_1(t))}{x_2(t)\, x_4(t)}$$

Elements of the Jacobian $\mathbf{D}_\alpha$

$$\mathbf{D_\alpha}(1, 1) = \frac{A\, a_4\, (b_1 - x_1(t))}{x_2(t)}$$

$$\mathbf{D_\alpha}(1, 2) = 0$$

$$\mathbf{D_\alpha}(1, 3) = 0$$

$$\mathbf{D_\alpha}(1, 4) = \frac{A\, (b_1 - x_1(t))\, (a_1 + 1)}{x_2(t)}$$

$$\mathbf{D_\alpha}(1, 5) = 0$$

$$\mathbf{D_\alpha}(2, 1) = \frac{A\, B\, a_4\, (b_1 - x_1(t))}{x_4(t)}$$

$$\mathbf{D}_\alpha(2,2) = 0$$

$$\mathbf{D}_\alpha(2,3) = 0$$

$$\mathbf{D}_\alpha(2,4) = \frac{A B a_1 \, (b_1 - x_1(t))}{x_4(t)}$$

$$\mathbf{D}_\alpha(2,5) = A$$

$$\mathbf{D}_\alpha(3,1) = \frac{A B a_4 x_5(t) \, (b_1 - x_1(t))}{Y x_2(t) x_4(t)}$$

$$\mathbf{D}_\alpha(3,2) = 0$$

$$\mathbf{D}_\alpha(3,3) = 0$$

$$\mathbf{D}_\alpha(3,4) = \frac{A}{x_2(t)} \left[q_s - x_3(t) + \frac{B a_1 x_5(t) \, (b_1 - x_1(t))}{Y x_4(t)} \right]$$

$$\mathbf{D}_\alpha(3,5) = 0$$

$$\mathbf{D}_\alpha(4,1) = -A a_4 \left[\frac{1}{x_2(t)} - \frac{X a_2}{x_4(t)} \right] (b_1 - x_1(t))$$

$$\mathbf{D}_\alpha(4,2) = \frac{A X a_1 a_4 \, (b_1 - x_1(t))}{x_4(t)}$$

$$\mathbf{D}_\alpha(4,3) = 0$$

$$\mathbf{D}_\alpha(4,4) = -A \left[\frac{a_1 + 1}{x_2(t)} - \frac{X a_1 a_2}{x_4(t)} \right] (b_1 - x_1(t))$$

$$\mathbf{D}_\alpha(4,5) = 0$$

$$\mathbf{D}_\alpha(5,1) = -\frac{A B H a_4}{x_4(t)} \left[\frac{x_5(t)}{x_2(t)} - \frac{H a_3}{M} \right] (b_1 - x_1(t))$$

$$\mathbf{D}_\alpha(5,2) = 0$$

$$\mathbf{D}_\alpha(5,3) = \frac{A B H a_1 a_4 \, (b_1 - x_1(t))}{M x_4(t)}$$

$$\mathbf{D}_\alpha(5,4) = -A \left[\frac{Y \, (b_3 - x_3(t))}{x_2(t)} + \frac{B H a_1}{x_4(t)} \left[\frac{x_5(t)}{x_2(t)} - \frac{H a_3}{M} \right] (b_1 - x_1(t)) \right]$$

$$\mathbf{D}_\alpha(5,5) = 0$$

Chapter 7
Lagrangian Tracer Dynamics

With the steady growth of the interest in ocean circulation systems and their impact on climate change, there has been a predictable increase in the number of tracer/drifter/buoy type ocean observing systems. There is a rich and growing literature on the development and testing of data assimilation technology to effectively utilize this new type of data set. This class of data assimilation has come to be known as Lagrangian data assimilation.

In one of the earlier studies, Carter (1989) examined the process of assimilating data from a set of 40 neutrally buoyant floats that followed the Gulf stream, which measured the location and depth collected three times a day, for 45 days. He combined this data with the nonlinear shallow water model using the well known extended Kalman filtering model (Lewis et al. 2006, Chaps. 27–29). Molcard et al. (2003) using a quasi-geostrophic reduced gravity model equations in a twin experiment set up generated circulation related data and developed an assimilation scheme that is based on the classical optimum interpolation (OI) technique (Lewis et al. 2006, Chap. 19). Again Özgökmen et al. (2003) examined the use of Kalman filter based method to assimilate drifter observations into an idealized layered primitive equation model.

By augmenting the state equations for a point vortex model with the drifter dynamics, Kuznetsov et al. (2003) examined the use of extended Kalman filter methodology to assimilate tracer data. Salman et al. (2008) recently explored the effectiveness of drifter deployment strategies using a nonlinear shallow water model with external wind forcing and combined it with the ensemble Kalman filter technology. Recently, Apte et al. (2008) describe a Bayesian approach to data assimilation using a simple linearized version of the shallow water model equations.

Our goal in this chapter is twofold: first is to explore the bifurcation exhibited by the tracer dynamics induced by the low/reduced order version of the shallow water model obtained using the standard Galerkin type projection method, and second is to demonstrate the applicability of the forward sensitivity method (FSM) (Lakshmivarahan and Lewis 2010) for assimilating tracer data.

© Springer International Publishing Switzerland 2017
S. Lakshmivarahan et al., *Forecast Error Correction using Dynamic Data Assimilation*, Springer Atmospheric Sciences, DOI 10.1007/978-3-319-39997-3_7

To this end, instead of relying on numerical methods, we first solve the resulting low order model which is a linear model equations in u, v and h given in Apte et al. (2008) in closed form. Using this explicit solution, we then express the flow field of the tracer dynamics as a sum of two parts—a time invariant nonlinear geostrophic mode, $\mathbf{f}(\mathbf{x}, u_0)$ depending on the geostrophic parameter u_0 and a time varying nonlinear part known as the inertial gravity mode, $g(\mathbf{x}, \boldsymbol{\alpha})$ depending on three parameters $\boldsymbol{\alpha} = (u_1(0), v_1(0), h_1(0))^T \in \mathbb{R}^3$. It turns out that the tracer dynamics controlled by the four parameters $(u_0, \alpha) \in \mathbb{R}^4$ exhibits complex behavior.

In Sect. 7.1, we derive the explicit expressions for the tracer dynamics which is a system of two first-order, coupled nonlinear, time varying ordinary differential equations. A complete catalog of all of the equilibria and their character is presented in Sect. 7.2. In Sect. 7.3 we examine the bifurcation of the tracer dynamics by succinctly summarizing the dependence on the four dimensional parameter set using a simple two dimensional characterization. The dynamics of evolution of the sensitivities of the solution is given in Sect. 7.4. The results of the data assimilation experiment using FSM are described in Sects. 7.5 and 7.6. Section 7.7 contains notes and references.

7.1 Shallow Water Model and Tracer Dynamics

Let $(x, y)^T \in \mathbb{R}^2$ denote the two dimensional space coordinates and $t \geq 0$ denote the time variable. Let $u(x, y, t)(v(x, y, t))$ denote the east-west (north-south) components of the velocity field at the spacial location $(x, y)^T$ and time t. Let $h(x, y, t)$ denote the height of the free surface of water measured above a pre-specified mean level. Let u_t, u_x, u_y denote the first partial derivatives of u with respect to t, x and y, respectively, and likewise for v and h.

The linear inviscid shallow water model equations are given by (Apte et al. 2008)

$$u_t = v - h_x,$$
$$v_t = -u - h_y, \qquad\qquad (7.1.1)$$
$$h_t = -u_x - u_y.$$

7.1.1 Low-Order Model (LOM)

Our analysis depends on the low-order counter part of the infinite dimensional model in (7.1.1) obtained by the application of the standard Galerkin type projection method. To this end, we first express u, v and h in the standard two dimensional truncated Fourier series consisting of only two terms given by

$$u(x, y, t) = -(2\pi l)u_0 \sin(2\pi Kx) \cos(2\pi Ly) + u_1(t) \cos(2\pi My),$$

$$v(x, y, t) = +(2\pi k)u_0 \cos(2\pi Kx) \sin(2\pi Ly) + v_1(t) \cos(2\pi My),$$

$$h(x, y, t) = \qquad u_0 \sin(2\pi Kx) \sin(2\pi Ly) + h_1(t) \sin(2\pi My). \quad (7.1.2)$$

The first time independent term with constant amplitude that is proportional to u_0 is called the *geostrophic mode* and the second time dependent terms with amplitudes $u_1(t), v_1(t), h_1(t)$ are called *inertial gravity modes*. In the following, we refer to u_0 as the *geostrophic parameter* and $u_1(0), v_1(0), h_1(0)$ as the *inertial parameters*. Without loss of generality and for simplicity in the exposition we only consider the case where $N = M = L = 1$ in (7.1.2).

Substituting (7.1.2) in (7.1.1), using the orthogonality property of the of standard trigonometric functions (Lewis et al. 2006, Appendix G) and simplifying, we obtain the following low-order model equations

$$\dot{u}_0 = 0, \qquad (7.1.3a)$$

$$\dot{\xi} = A\xi, \qquad (7.1.3b)$$

where $\xi = (u_1(t), v_1(t), h_1(t))^T \in \mathbb{R}$, and $A \in \mathbb{R}^{3\times 3}$ is a *skew-symmetric* matrix[1] given by

$$A = \begin{bmatrix} 0 & 1 & 0 \\ -1 & 0 & -a \\ 0 & a & 0 \end{bmatrix}, \qquad (7.1.4)$$

with $a = 2\pi$ and $\dot{u}_0$ and $\dot{\xi}$ denote the time derivatives of u_0 and ξ respectively.

From (7.1.3a) it is evident that u_0 is a constant, and (7.1.3b) conserves energy as shown below.

Property 7.1.1. Let $E(t) = \frac{1}{2} \|\xi^T(t)\|^2$ denote the energy. Then (7.1.3b) conserves energy $E(t)$, that is, $\dot{E}(t) = 0$.

Proof. For,

$$\dot{E} = \xi^T(t)\,\dot{\xi}(t) = \xi^T(t)\,A\,\xi(t) = 0, \qquad (7.1.5)$$

since A is a *skew-symmetric* matrix. Stated in other words, the solution $\xi(t)$ of the linear system (7.1.3b) always lies on the surface of a sphere centered at the origin with radius given by

$$E(0)^{1/2} = \left[u_1^2(0) + v_1^2(0) + h_1^2(0)\right]^{1/2} = E(t)^{1/2}. \qquad (7.1.6)$$

---

[1] A real matrix $A \in \mathbb{R}^{n\times n}$ is symmetric if $A^T = A$ and is skew-symmetric if $A^T = -A$.

7.1.2 Solution of LOM

Clearly, the matrix A in (7.1.4) is singular with eigenvalues given by

$$\lambda_1 = i\lambda, \lambda_2 = -i\lambda, \lambda_3 = 0, \text{ where } \lambda = \sqrt{1 + a^2}, \text{ and } a = 2\pi. \tag{7.1.7}$$

Accordingly, the Jordan canonical form (Meyer 2000) of A is given by

$$\mathbf{A} = \mathbf{V}\mathbf{\Lambda}\mathbf{V}^{-1}, \tag{7.1.8}$$

where

$$\mathbf{\Lambda} = \begin{bmatrix} 0 & \lambda & 0 \\ -\lambda & 0 & 0 \\ 0 & 0 & 0 \end{bmatrix}, \tag{7.1.9}$$

and the matrix V of corresponding eigenvectors is given by

$$\mathbf{V} = \begin{bmatrix} 0 & -\frac{1}{\lambda} & \frac{a}{\lambda} \\ 1 & 0 & 0 \\ 0 & -\frac{a}{\lambda} & -\frac{1}{\lambda} \end{bmatrix}. \tag{7.1.10}$$

where

$$c = \cos(\lambda t), \text{ and } s = \sin(\lambda t). \tag{7.1.11}$$

It can be verified that $\mathbf{V}^{-1} = \mathbf{V}^T$, that is, $\mathbf{V}$ is an orthogonal matrix. The solution $\boldsymbol{\xi}(t)$ of (7.1.3b) is then given by

$$\boldsymbol{\xi}(t) = e^{\mathbf{A}t}\boldsymbol{\xi}(0) = e^{(\mathbf{V}\mathbf{\Lambda}\mathbf{V})t}\boldsymbol{\xi}(0) = \left[\mathbf{V}e^{(\mathbf{\Lambda}\mathbf{V})}\right]\boldsymbol{\xi}(0), \tag{7.1.12a}$$

where

$$e^{\mathbf{\Lambda}t} = \begin{bmatrix} c & s & 0 \\ -s & c & 0 \\ 0 & 0 & 1 \end{bmatrix}. \tag{7.1.12b}$$

Substituting (7.1.12b) in (7.1.12a) and simplifying we obtain the components of the solution $\boldsymbol{\xi}(t)$ as:

$$u_1(t) = \frac{c + a^2}{\lambda^2} u_1(0) + \frac{s}{\lambda} v_1(0) + \frac{ac - a}{\lambda^2} h_1(0),$$

$$v_1(t) = \frac{-s}{\lambda} u_1(0) + c\, v_1(0) + \frac{-as}{\lambda} h_1(0),$$

$$h_1(t) = \frac{ac - a}{\lambda^2} u_1(0) + \frac{as}{\lambda} v_1(0) + \frac{1 + ca^2}{\lambda^2} h_1(0). \tag{7.1.13a}$$

After rearranging,

$$u_1(t) = \frac{a}{\lambda^2}[au_1(0) - h_1(0)] + \frac{1}{\lambda^2}[u_1(0) + ah_1(0)]c + \frac{v_1(0)}{\lambda}s,$$

$$v_1(t) = v_1(0)c - \frac{1}{\lambda}[ah_1(0) + u_1(0)]s,$$

$$h_1(t) = \frac{1}{\lambda^2}[h_1(0) - au_1(0)] + \frac{a}{\lambda^2}[u_1(0) + ah_1(0)]c + \frac{av_1(0)}{\lambda}s.$$

$$(7.1.13b)$$

Since this solution $\xi(t)$ lies on a sphere, it is immediate that $u_1(t)$, $v_1(t)$, and $h_1(t)$ are not linearly independent.

7.1.3 Tracer Dynamics

Let $\mathbf{x}(t) = (x(t), y(t))^T \in \mathbb{R}^2$ denote the position of the tracer particle floating on the free surface of the water. The dynamics of motion of this tracer is then given by

$$\begin{aligned} \dot{x} &= u(x, y, t), \\ \dot{y} &= v(x, y, t), \end{aligned} \qquad (7.1.14)$$

where the right hand side of (7.1.14) is obtained by substituting (7.1.13) in (7.1.2) with u_0, a constant. For later reference, define the velocity components

$$\begin{aligned} f_1(x, y) &= -2\pi u_0 \sin(2\pi x) \cos(2\pi y), \\ f_2(x, y) &= 2\pi u_0 \cos(2\pi x) \sin(2\pi y), \end{aligned} \qquad (7.1.15)$$

and

$$\begin{aligned} g_1(x, y, t) &= u_1(t) \cos(2\pi My), \\ g_2(x, y, t) &= v_1(t) \cos(2\pi My). \end{aligned} \qquad (7.1.16)$$

Let $\mathbf{f} = (f_1, f_2)^T \in \mathbb{R}^2$ and $\mathbf{g} = (g_1, g_2)^T \in \mathbb{R}^2$. Then, (7.1.14) can be rewritten in the vector form as

$$\dot{\mathbf{x}}(t) = \mathbf{f}(\mathbf{x}, u_0) + \mathbf{g}(\mathbf{x}, \boldsymbol{\alpha}, t), \qquad (7.1.17)$$

where $u_1(t), v_1(t), h_1(t)$ are given in (7.1.13) and recall that $\boldsymbol{\alpha} = (u_1(0), v_1(0), h_1(0)) \in \mathbb{R}^3$. Clearly, the first component f of the vector field depends on the geostrophic parameter u_0 and the second component g depends on the inertial parameters $\boldsymbol{\alpha} \in \mathbb{R}^3$. Our goal in this paper is to characterize and catalog the behavior of the solution of (7.1.17) as the geostrophic and inertial parameters are varied. It turns out this class of nonlinear time varying dynamics exhibits many interesting bifurcations (Kuznetsov 2013) as the four parameters are varied in $\mathbb{R}^4$.

7.2 Analysis of Equilibria of Tracer Dynamics

It is convenient to divide the analysis into three cases.

7.2.1 Case 1: Equilibria in Geostrophic Mode

By setting the inertial parameter $\alpha = 0$, we obtain the tracer dynamics given by the nonlinear autonomous system

$$\dot{\mathbf{x}} = \mathbf{f}(\mathbf{x}, u_0) \tag{7.2.1}$$

controlled by the geostrophic parameter u_0. From (7.1.15), the Jacobian of the flow field $\mathbf{f}(\mathbf{x}, u_0)$ is given by

$$\mathbf{D_x(f)} = -4\pi^2 u_0 \begin{bmatrix} \cos(2\pi x)\cos(2\pi y) & -\sin(2\pi x)\sin(2\pi y) \\ \sin(2\pi x)\sin(2\pi y) & -\cos(2\pi x)\cos(2\pi y) \end{bmatrix}. \tag{7.2.2}$$

We consider two types of equilibria for (7.2.1).

7.2.1.1 Type 1 Equilibria

From (7.2.1) and (7.1.15), it follows that $\mathbf{f}(\mathbf{x}, u_0) = 0$ when $sin(2\pi x) = 0 = sin(2\pi y)$, that is, when

$$x = y = \pm\frac{k}{2} \text{ for } k = 0, 1, 2, \ldots \tag{7.2.3}$$

which is independent of u_0. The Jacobian (7.2.2) at these equilibria becomes

$$\mathbf{D_x(f)} = -4\pi^2 u_0 \begin{bmatrix} 1 & 0 \\ 0 & -1 \end{bmatrix}, \tag{7.2.4}$$

whose eigenvalues are given by $\lambda_1 = 4\pi^2 u_0$ and $\lambda_2 = -4\pi^2 u_0$, with $\mathbf{e}_1 = (1, 0)^T$ and $\mathbf{e}_2 = (0, 1)^T$ as their corresponding eigenvectors. Clearly, this family of equilibria correspond to a *saddle* (Kuznetsov 2013) for all non-zero values of the geostrophic parameter u_0. In a small neighborhood around the equilibria in (7.2.3), the system (7.2.1) is equivalent to a linear dynamics given by

$$\dot{\eta} = \mathbf{D_x(f)}\eta, \tag{7.2.5}$$

with $\mathbf{D_x(f)}$ in (7.2.4) and $\eta(t) = (\eta_1(t), \eta_2(t))^T$.
The solution of (7.2.5) is given by

$$\eta_1(t) = \eta_1(0)\,e^{-(4\pi^2 u_0)t} \text{ and } \eta_2(t) = \eta_2(0)\,e^{(4\pi^2 u_0)t}. \tag{7.2.6}$$

7.2.1.2   Type 2 Equilibria

From (7.2.1) and (7.1.15), it again follows that $\mathbf{f}(\mathbf{x}, u_0) = 0$ when $\cos(2\pi x) = 0 = \cos(2\pi y)$, that is, when

$$x = y = \pm \frac{2k + 1}{4} \quad \text{for } k = 0, 1, 2, \ldots \tag{7.2.7}$$

which is also independent of u_0. The Jacobian (7.2.2) at these equilibria becomes

$$\mathbf{D_x}(\mathbf{f}) = -4\pi^2 u_0 \begin{bmatrix} 0 & -1 \\ 1 & 0 \end{bmatrix}, \tag{7.2.8}$$

whose eigenvalues are purely imaginary and are given by $\lambda_1 = i\,4\pi^2 u_0$ and $\lambda_2 = -i\,4\pi^2 u_0$, with any pair of unit, orthogonal vectors $e_1 = \frac{1}{\sqrt{a^2+b^2}}(a, b)^T$ and $e_2 = \frac{1}{\sqrt{a^2+b^2}}(b, -a)^T$ as corresponding eigenvectors, where a and b are arbitrary real constants. This family of equilibria correspond to a *center* (Kuznetsov 2013) for all non-zero values of the geostrophic parameter u_0. In a small neighborhood around this equilibria, Eq. (7.2.1) is equivalent to a linear dynamics

$$\dot{\eta} = \mathbf{D_x}(\mathbf{f})\eta, \tag{7.2.9}$$

with $\mathbf{D_x}(\mathbf{f})$ in (7.2.8). The solution of (7.2.9) is then given by

$$\begin{bmatrix} \eta_1(t) \\ \eta_2(t) \end{bmatrix} = \begin{bmatrix} \cos(bt) & \sin(bt) \\ -\sin(bt) & \cos(bt) \end{bmatrix} \begin{bmatrix} \eta_1(0) \\ \eta_2(0) \end{bmatrix}, \quad \text{where } b = 4\pi^2 u_0. \tag{7.2.10}$$

It turns out that Eq. (7.2.1) with the field $\mathbf{f}(\mathbf{x}, u_0)$ given in (7.1.15) can indeed be solved in closed form. It can be verified that (with $a = 2\pi$), (7.2.1) is equivalent to

$$\frac{dy}{dx} = -\frac{\tan(ay)}{\tan(ax)}. \tag{7.2.11}$$

Expressing (7.2.11) as

$$\frac{dy}{\tan(ay)} = -\frac{dx}{\tan(ax)} \tag{7.2.12}$$

and integrating, the solution is given by

$$\log(\sin(ax)\sin(ay)) = c, \tag{7.2.13}$$

where c is the constant of integration. If (x_0, y_0) is the initial position of the tracer, from (7.2.13) the solution of (7.2.1) can be expressed as

$$\sin(ax)\sin(ay) = \sin(ax_0)\sin(ay_0) \tag{7.2.14}$$

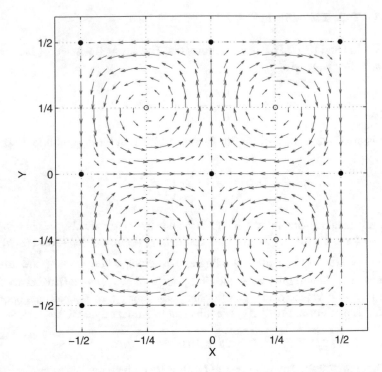

Fig. 7.1 A display of equilibria of *Type 1—filled circles* and *Type 2—unfilled circles* along with the velocity field around them. *Filled circles* are saddle points and *unfilled circles* are centers. The field plot around these equilibria corresponds to $f(x, y)$ in (7.1.15) with $u_0 = 1.0, u_1(0) = v_1(0) = h_1(0) = 0$ and time $t = 0$

Figure 7.1 is an illustration of the relative disposition *types 1* and *2* equilibria along with the vector field around them when $u_0 = 1.0$. Phase plots using (7.2.14) for various choices of (x_0, y_0) are given in Fig. 7.2.

7.2.2   Case 2: Equilibria in Inertia-Gravity Mode

By setting the geostrophic parameter $u_0 = 0$ and ensuring that $\alpha \neq 0$, we obtain the second special case for the tracer dynamics given by the nonlinear nonautonomous system

$$\dot{\mathbf{x}} = \mathbf{g}(\mathbf{x}, \alpha, t), \tag{7.2.15}$$

whose behavior is controlled by $\alpha \in \mathbb{R}^3$. Referring to (7.1.16), the Jacobian of this flow field is given by

$$\mathbf{D_x}(\mathbf{g}) = 2\pi \begin{bmatrix} 0 & -u_1(t) \sin(2\pi y) \\ 0 & -v_1(t) \sin(2\pi y) \end{bmatrix}. \tag{7.2.16}$$

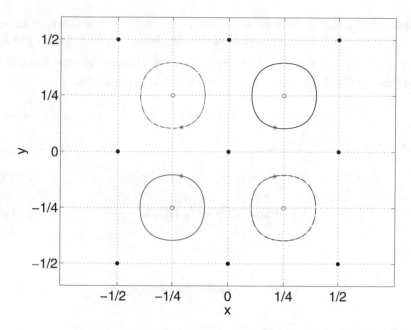

Fig. 7.2 A display of trajectories of the pure geostrophic dynamics $\dot{x} = f(x, u_0)$ in (7.2.1). The values of $(x_0, y_0) = [(0.209, 0.109), (-0.209, 0.109), (0.209, -0.109), (-0.209, -0.109)]$ are indicated by *asterisks*

From (7.1.16) it is immediate that $\mathbf{g}(\mathbf{x}, \alpha, t) = 0$, when $\cos(2\pi y) = 0$. Thus, the equilibria for (7.2.15) are given by

$$x \text{ arbitrary, and } y = \pm\frac{2k+1}{4}, \quad \text{for } k = 0, 1, 2, \ldots \tag{7.2.17}$$

It is convenient to consider two types of equilibria.

7.2.2.1 Type a: Equilibria in the Upper Half Plane

This type of equilibria is given by

$$x \text{ arbitrary, and } y = \frac{2k+1}{4}, k = 0, 1, 2, \ldots \tag{7.2.18}$$

The Jacobian (7.2.16) evaluated at these equilibria becomes

$$\mathbf{D}_{\mathbf{x}}(\mathbf{g}) = 2\pi \begin{bmatrix} 0 & -u_1(t) \\ 0 & -v_1(t) \end{bmatrix}, \tag{7.2.19}$$

whose eigenvalues are given by $\lambda_1 = -2\pi v_1(t)$ and $\lambda_2 = 0$, where $v_1(t)$ depends on α through (7.1.13). It can be verified that the eigenvector corresponding to λ_1 is $\mathbf{e}_1 = \left(\frac{u_1(t)}{v_1(t)}, 1\right)^T$ and that corresponding to λ_2 is any arbitrary non-zero vector in $\mathbb{R}^2$. Dynamics in (7.2.15) around the equilibrium (7.2.18) is equivalent to

$$\dot{\eta} = \mathbf{D}_\mathbf{x}(\mathbf{g})\eta, \qquad (7.2.20)$$

with $\mathbf{D}_\mathbf{x}(\mathbf{g})$ in (7.2.19). Clearly,

$$\eta_2(t) = \eta_2(0)\, e^{-2\pi\Psi(t)}, \qquad (7.2.21a)$$

$$\text{where } \Psi(t) = \int_0^t v_1(s)ds. \qquad (7.2.21b)$$

Hence,

$$\eta_1(t) = -2\pi\eta_2(0) \int_0^t u_1(s)\eta_2(s)ds. \qquad (7.2.22)$$

7.2.2.2 Type b: Equilibria in the Lower Half Plane

This type of equilibria is given by

$$x \text{ arbitrary, and } y = -\frac{2k+1}{4}, \text{ for } k = 0, 1, 2, \dots \qquad (7.2.23)$$

The Jacobian (7.2.16) evaluated at these equilibria becomes

$$\mathbf{D}_\mathbf{x}(\mathbf{g}) = 2\pi \begin{bmatrix} 0 & u_1(t) \\ 0 & v_1(t) \end{bmatrix}, \qquad (7.2.24)$$

whose eigenvalues are given by $\lambda_1 = 2\pi v_1(t)$ and $\lambda_2 = 0$, whose eigenvectors are the same as in *Type a*. Again the nonlinear equation (7.2.15) is equivalent to

$$\dot{\eta} = \mathbf{D}_\mathbf{x}(\mathbf{g})\eta \qquad (7.2.25)$$

around the equilibria in (7.2.23) with $\mathbf{D}_\mathbf{x}(\mathbf{g})$ in (7.2.24). The solution of (7.2.25) is given by

$$\eta_2(t) = \eta_2(0)\, e^{2\pi\Psi(t)}$$

$$\eta_1(t) = 2\pi\eta_1(0) \int_0^t u_1(s)\eta_2(s)ds \qquad (7.2.26)$$

Again it turns out that we can solve (7.2.15) from (7.1.16). The second equation in (7.2.15) becomes

$$\frac{dy}{\cos{(ay)}} = v_1(t)dt, \tag{7.2.27}$$

which on integrating becomes

$$\frac{1}{a}\ln\left(\tan\left(\frac{\pi}{4} + \frac{ay}{2}\right)\right) = c + \Psi(t), \tag{7.2.28}$$

where $\Psi(t)$ is given in (7.2.21b). If (x_0, y_0) is the initial position of the tracer, the solution of (7.2.28) becomes

$$\tan\left(\frac{\pi}{4} + \frac{ay}{2}\right) = e^{2\pi\Psi(t)}\tan\left(\frac{\pi}{4} + \frac{ay_o}{2}\right), \quad \text{or} \tag{7.2.29a}$$

$$y(t) = \frac{2}{a}\left[\arctan\left(e^{2\pi\Psi(t)}\tan\left(\frac{\pi}{4} + \frac{ay_o}{2}\right)\right) - \frac{\pi}{4}\right]. \tag{7.2.29b}$$

The first component of the solution of (7.2.15) is given by

$$x(t) = x(0) + \int_0^t u_1(s)\cos{(2\pi y(s))}ds. \tag{7.2.30}$$

An illustration of the relative disposition of the equilibria and the field corresponding to (7.2.15) are given in Fig. 7.3. Figure 7.4 contains the phase plot $(x(t), y(t))$ obtained from (7.2.29) and (7.2.30) for different initial points.

7.2.3 Case 3: Equilibria in General Case

When $u_0 \neq 0$ and $\alpha \neq 0$, we obtain the interesting general case where the geostrophic and the inertial gravitational modes exert their own influence. Depending on the relative strength of these component fields, we can obtain a variety of behavior of the tracer dynamics

$$\dot{\mathbf{x}} = \mathbf{f}(\mathbf{x}, u_0) + \mathbf{g}(\mathbf{x}, \alpha, t). \tag{7.2.31}$$

The Jacobian of this flow field is given by

$$\mathbf{D_x}(\mathbf{f} + \mathbf{g}) = \mathbf{D_x}(\mathbf{f}) + \mathbf{D_x}(\mathbf{g})$$

$$= 2\pi\begin{bmatrix} -2\pi u_0 \cos(2\pi x)\cos(2\pi y) & 2\pi u_0 \sin(2\pi x)\sin(2\pi y) - u_1(t)\sin(2\pi y) \\ -2\pi u_0 \sin(2\pi x)\sin(2\pi y) & 2\pi u_0 \cos(2\pi x)\cos(2\pi y) - v_1(t)\sin(2\pi y) \end{bmatrix}. \tag{7.2.32}$$

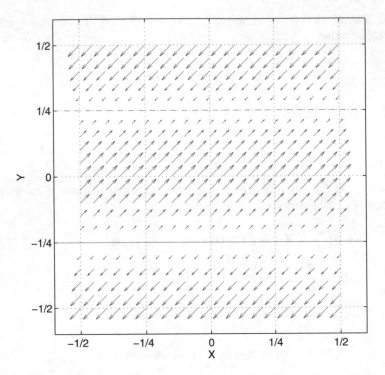

Fig. 7.3 A display of equilibria of *Type a—dashed line* and *Type b—solid line*, and the flow field around them. The plot of the time varying vector field around these equilibria corresponding to $g(x, \alpha, t)$ in (7.2.17) at time $t = 0$ and for values of corresponding parameters $u0 = 0.0, u1(0) = v1(0) = h1(0) = 1$ and time $t = 0$

It can be verified that $\mathbf{f}(\mathbf{x}, u_0) + \mathbf{g}(\mathbf{x}, \boldsymbol{\alpha}) = 0$ when

$$x = y = \pm\frac{2k + 1}{4}, \quad \text{for } k = 0, 1, 2, \ldots \tag{7.2.33}$$

To simplify the analysis we consider four types of equilibria.

7.2.3.1   Type A: Equilibria in the First Quadrant

$$x = y = \frac{2k + 1}{4}, \quad \text{for } k = 0, 1, 2, \ldots$$

In this case, the Jacobian (7.2.32) becomes

$$\mathbf{D}_{\mathbf{x}}(\mathbf{f} + \mathbf{g}) = 2\pi \begin{bmatrix} 0 & 2\pi u_0 - u_1(t) \\ -2\pi u_0 & -v_1(t) \end{bmatrix}, \tag{7.2.34}$$

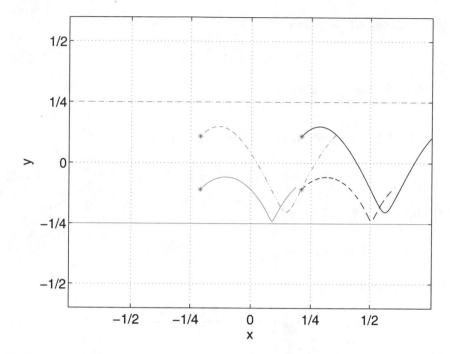

Fig. 7.4 A display of trajectories of the pure inertial gravity modes $\dot{x} = g(x, \alpha, t)$ in (7.2.1) The starting points of various trajectories $(x_0, y_0) = [(0.209, 0.109), (-0.209, 0.109), (0.209, -0.109), (-0.209, -0.109)]$ are shown by *asterisks*

whose eigenvalues are given by

$$\lambda_{1,2}(t) = \frac{-v_1(t) \pm \sqrt{\Delta_1(t)}}{2}, \tag{7.2.35}$$

where

$$\Delta_1(t) = (v_1(t))^2 - 8\pi u_0 \left(2\pi u_0 - u_1(t)\right). \tag{7.2.36}$$

7.2.3.2 Type B: Equilibria in the Fourth Quadrant

$$x = -y = \frac{2k + 1}{4}, \quad \text{for } k = 0, 1, 2, \ldots$$

In this case, the Jacobian (7.2.32) takes the form

$$\mathbf{D_x(f + g)} = 2\pi \begin{bmatrix} 0 & -2\pi u_0 + u_1(t) \\ 2\pi u_0 & v_1(t) \end{bmatrix}, \tag{7.2.37}$$

whose eigenvalues are given by

$$\lambda_{1,2}(t) = \frac{v_1(t) \pm \sqrt{\Delta_1(t)}}{2}, \tag{7.2.38}$$

where Δ_1 is given in (7.2.36).

7.2.3.3   Type C: Equilibria in the Second Quadrant

$$x = -y = -\frac{2k+1}{4}, \quad \text{for } k = 0, 1, 2, \ldots \tag{7.2.39}$$

The Jacobian (7.2.32) becomes

$$\mathbf{D_x}(\mathbf{f} + \mathbf{g}) = 2\pi \begin{bmatrix} 0 & -2\pi u_0 - u_1(t) \\ 2\pi u_0 & -v_1(t) \end{bmatrix}, \tag{7.2.40}$$

whose eigenvalues are given by

$$\lambda_{1,2}(t) = \frac{-v_1(t) \pm \sqrt{\Delta_2(t)}}{2}, \tag{7.2.41}$$

where

$$\Delta_2(t) = (v_1(t))^2 - 8\pi u_0 \left(2\pi u_0 + u_1(t)\right). \tag{7.2.42}$$

7.2.3.4   Type D: Equilibria in the Fourth Quadrant

$$x = y = -\frac{2k+1}{4}, \quad \text{for } k = 0, 1, 2, \ldots$$

The Jacobian (7.2.32) at this equilibrium becomes

$$\mathbf{D_x}(\mathbf{g}) = 2\pi \begin{bmatrix} 0 & 2\pi u_0 + u_1(t) \\ 2\pi u_0 & v_1(t) \end{bmatrix}, \tag{7.2.43}$$

whose eigenvalues are given by

$$\lambda_{1,2}(t) = \frac{v_1(t) \pm \sqrt{\Delta_2(t)}}{2}, \tag{7.2.44}$$

where Δ_2 is given in (7.2.42).

The character of the equilibria of *Types A* and *B* depends on the sign of $\Delta_1(t)$ in (7.2.36) and that of the equilibria of *Types C* and *D* depends on the sign of $\Delta_2(t)$ in (7.2.42).

7.2.4 Conditions for the Sign Definiteness of $\Delta_i(t)$

From (7.2.36) and (7.2.42), since $v_1^2(t) \geq 0$, a necessary and sufficient condition for $\Delta_i(t) > 0$ for all $t \geq 0$ is given by

$$\Delta_1(t) \geq 0 \quad \text{when} \quad u_1(t) \geq 2\pi u_0 \qquad \text{if} \quad u_0 > 0, \tag{7.2.45a}$$

$$u_1(t) \leq 2\pi |u_0| \qquad \text{if} \quad u_0 < 0, \tag{7.2.45b}$$

and

$$\Delta_2(t) \geq 0 \quad \text{when} \quad u_1(t) \leq -2\pi u_0 \qquad \text{if} \quad u_0 > 0, \tag{7.2.46a}$$

$$u_1(t) \geq 2\pi |u_0| \qquad \text{if} \quad u_0 < 0. \tag{7.2.46b}$$

Since $u_1(t)$ depends on α, for a given value of (u_0, α), inequalities (7.2.45a) and (7.2.46a) cannot hold simultaneously. Similarly, (7.2.45b) and (7.2.46b) cannot also hold simultaneously. Thus, for a given (u_0, α), if $\Delta_1(t) \geq 0$ then $\Delta_2(t) \leq 0$, and vice versa. This in turn implies that if the eigenvalues of *Types A* and *B* equilibria are real, then those of *Type C* and *D* are complex conjugates and vice versa. To further understand the nature of the real eigenvalues, consider the case when $u_0 > 0$ and $u_1(t) \geq 2\pi u_0$. For this choice,

$$\Delta_1(t) = (v_1(t))^2 - 8\pi u_0(2\pi u_0 - u_1(t))$$
$$= (v_1(t))^2 + 8\pi u_0(u_1(t) - 2\pi u_0). \tag{7.2.47}$$

Hence $\sqrt{\Delta_1(t)} > v_1(t)$ and the two eigenvalues are such that

$$\lambda_1(t) = \frac{-v_1(t) + \sqrt{\Delta_1(t)}}{2} \quad \text{and} \quad \lambda_2(t) = \frac{-v_1(t) - \sqrt{\Delta_1(t)}}{2}. \tag{7.2.48}$$

This is, when $u_0 > 0$ and $u_1(t) > 2\pi u_0$, the *Types A* and *B* equilibria are *saddle points* while *Types C* and *D* are *centers*. Stated in other words, *Types A* and *B*, and *Types C* and *D* have a mutually complementary character. A similar argument carries over to the case when $u_0 < 0$.

A display of these four types of equilibria along with the flow field around them is given in Fig. 7.5. Some sample solutions of (7.2.31) obtained numerically are also given in Fig. 7.6.

7.3 Analysis of Bifurcation

In Sect. 7.2 we have cataloged the properties of the set of all equilibria of the tracer dynamics in different regions of the four dimensional parameter space, $\mathbb{R}^4$ containing (u_0, α) namely, $u_0 \neq 0$ and $\alpha = 0$ in Case 1, $u_0 = 0$ and $\alpha \neq 0$ in

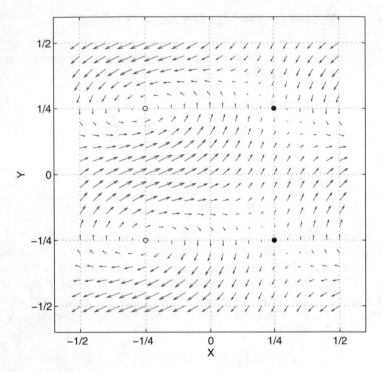

Fig. 7.5 A display of equilibria of *Type A* (1/4, 1/4) and *B* (1/4, −1/4)—*solid circles* and *Type C* (−1/4, 1/4) and *D* (−1/4, −1/4)—*empty circles*, and the flow field around them. A snapshot of the time varying vector field given by (7.2.31) at time $t = 0$ where $u_0 = 1.0$, $u_1(0) = 9.4248$, $v_1(0) = \lambda$ and $h_1(0) = -1.5$

Case 2, $u_0 \neq 0$ and $\alpha \neq 0$ in Case 3. In this section we examine the transition between the equilibria as the parameters are varied continuously.

To this end, we start by translating the conditions for the positive definiteness of $\Delta_i(t)$ in [(7.2.45a), (7.2.45b)] and [(7.2.46a), (7.2.46b)] directly in term of (u_0, α) using (7.1.13). Define a new set of parameters U, H and V through a linear transformation suggested by the expressions for $u_1(t)$ in (7.1.13) as

$$
\begin{aligned}
U &= \frac{a^2}{\lambda^2} u_1(0) - \frac{a}{\lambda^2} h_1(0), \\
H &= \frac{1}{\lambda^2} u_1(0) + \frac{a}{\lambda^2} h_1(0),
\end{aligned}
\tag{7.3.1}
$$

$$
V = \frac{1}{\lambda} v_1(0).
\tag{7.3.2}
$$

Rewriting (7.3.1) in matrix form, it can be verified that

$$
\begin{bmatrix} U \\ H \end{bmatrix} = \frac{1}{\lambda^2} \begin{bmatrix} a^2 & -a \\ 1 & a \end{bmatrix} \begin{bmatrix} u_1(0) \\ h_1(0) \end{bmatrix}.
\tag{7.3.3}
$$

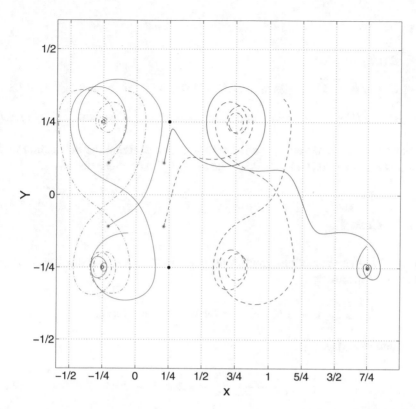

Fig. 7.6 A display of trajectories of (7.2.31) are given where the starting points $(x_0, y_0) = [(0.209, 0.109), (-0.209, 0.109), (0.209, -0.109), (-0.209, -0.109)]$ are indicated by *asterisks*

The 2×2 matrix in (7.3.3) is non-singular. Hence (7.3.1), (7.3.2) define an invertible linear transformation between $(u_1(0), v_1(0), h_1(0))^T \in \mathbb{R}^3$ to $(U, V, H) \in \mathbb{R}^3$.

In terms of these new variables, $u_1(t)$ in (7.1.13) now becomes

$$u_1(t) = U + H \cos(\lambda t) + V \sin(\lambda t). \tag{7.3.4}$$

Now invoking the results from section "Bounds on $u_1(t)$ in (7.1.13)" in Appendix, we obtain a uniform (in time) lower and upper bound on $u_1(t)$ given by

$$U - \sqrt{H^2 + V^2} \le u_1(t) \le U + \sqrt{H^2 + V^2}. \tag{7.3.5}$$

Using these bounds, we now translate the conditions for sign definiteness of $\Delta_i(t)$ in (7.2.45)–(7.2.46) in terms of these new set of parameters. It can be verified that

$\Delta_1(t) \ge 0$ when

$$U - \sqrt{H^2 + V^2} \ge 2\pi u_0 \text{ and } U \ge 2\pi u_0 \qquad \text{if } u_0 > 0, \tag{7.3.6a}$$

$$U + \sqrt{H^2 + V^2} \le -2\pi |u_0| \text{ and } U \le -2\pi |u_0| \quad \text{if } u_0 < 0, \tag{7.3.6b}$$

and

$\Delta_2(t) \geq 0$ when

$$U + \sqrt{H^2 + V^2} \leq -2\pi u_0 \text{ and } U \leq 2\pi u_0 \qquad \text{if } u_0 > 0, \qquad (7.3.7a)$$

$$U - \sqrt{H^2 + V^2} \geq 2\pi |u_0| \text{ and } U \geq 2\pi |u_0| \qquad \text{if } u_0 < 0. \qquad (7.3.7b)$$

Without the loss of generality, in the following we examine the bifurcation in the new parameter space $(U, V, H) \in \mathbb{R}^3$ and $u_0 \in \mathbb{R}$. We consider two cases.

7.3.1 Case 1

To visualize these conditions graphically, consider first the condition for $\Delta_1(t) \geq 0$ where $u_0 > 0$, namely,

$$U - \sqrt{H^2 + V^2} \geq 2\pi u_0, \text{ and } U \geq 2\pi u_0. \qquad (7.3.8)$$

Rewriting (7.3.8) as

$$\frac{(U - 2\pi u_0)^2}{V^2} - \frac{H^2}{V^2} \geq 1 \text{ and } U \geq 2\pi u_0, \qquad (7.3.9)$$

and referring to section "Definition and Properties of Standard Hyperbola" in Appendix, it turns out that (7.3.9) with the equality sign indeed represents the equation for a standard hyperbola in the $U - H$ (two dimensional) plane with the following key characteristics: the hyperbola is centered at $(2\pi u_0, 0)$, its semi-major and semi-minor axes are equal and are given by V; its eccentricity is $e = \sqrt{2}$ and the slopes of its asymptotes are ± 1. Since $U \geq 2\pi u_0$, we only have to consider the right branch of hyperbola.

Notice that the geostrophic parameter u_0 only affects the location of the center of the hyperbola and $v_1(0)$ (through V) only affects the length of the semi-axes. It is interesting to note that while the parameter space is really four dimensional, we can succinctly represent the effects of all the parameters using the standard hyperbola in two dimensions. Refer to Fig. 7.7 for an illustration.

Now consider the case for $\Delta_2(t) \geq 0$ when $u_0 > 0$, namely

$$U + \sqrt{H^2 + V^2} \leq -2\pi u_0, \text{ and } U \leq -2\pi u_0. \qquad (7.3.10)$$

Rewriting (7.3.10) as

$$\frac{(U + 2\pi u_0)^2}{V^2} - \frac{H^2}{V^2} \leq 1, \text{ and } U \leq -2\pi u_0, \qquad (7.3.11)$$

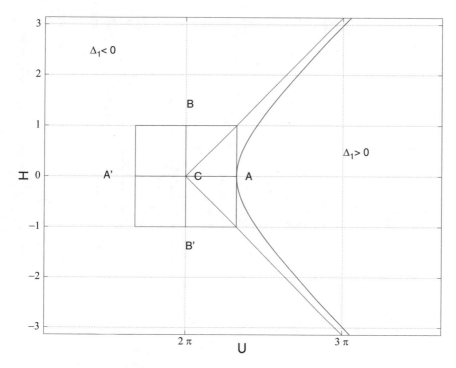

Fig. 7.7 Hyperbola corresponding to (7.3.9) C in the center at $(2\pi u_0, 0)$ with $u_0 = 1$. Let $v_1(0) = 1$ and the semi axes $AC = AC' = BC = 1$. The asymptotes have slope ± 1

it can be verified that (7.3.11) with equality sign denotes a hyperbola which shares all the characteristics of the one described above but centered $(-2\pi u_0, 0)$. Refer to Fig. 7.7 for an illustration. Again, in view of the constraint $U \leq 2\pi u_0$, we only have to consider the left branch of hyperbola.

The complete characterization of the four types of equilibria in Case 3 in Sect. 7.2.3 is obtained by superimposing two systems of hyperbolas in Figs. 7.7 and 7.8. This combined system along with the complete characteristic of various regions are given in Fig. 7.9. When $u_0 \rightarrow 0$, the center of the hyperbolas move towards the origin of the $U - H$ plane. Similarly, when $v_1(0) \rightarrow 0$, while the eccentricity $e = \sqrt{2}$ the slopes of the asymptotes remain constant at ± 1, the lengths of the semi axes shrink to zero.

7.3.2 Case 2

From (7.3.6) the condition for $\Delta_1(t) \geq 0$, when $u_0 \leq 0$, becomes

$$\frac{(U + 2\pi |u_0|)^2}{V^2} - \frac{H^2}{V^2} \leq 1, \quad \text{and} \quad U \leq -2\pi |u_0|. \tag{7.3.12}$$

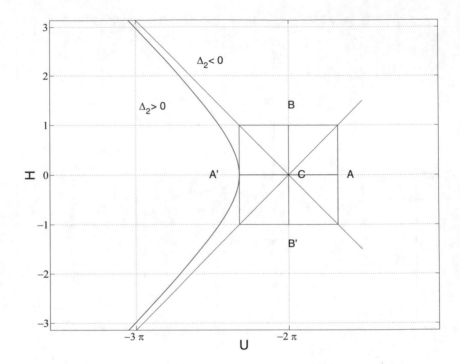

Fig. 7.8 Hyperbola corresponding to (7.3.11) C in the center at $(-2\pi u_0, 0)$ with $u_0 = 1$ Let $v_1(0) = 1$ and the semi axes $CA = CA' = CB = CB' = 1$. The asymptotes have slope ± 1

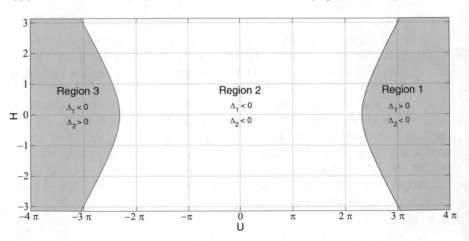

Fig. 7.9 The combined system of hyperbolas from Figs. 7.7 and 7.8. Regions corresponding to different signs of $\Delta_1(t)$ and $\Delta_2(t)$ are shown. Points on the hyperbola are the bifurcation points

and from (7.3.7) the conditions for $\Delta_2(t) \geq 0$, and $u_0 \leq 0$, becomes

$$\frac{(U - 2\pi |u_0|)^2}{V^2} - \frac{H^2}{V^2} \geq 1, \quad \text{and} \quad U \leq -2\pi |u_0|. \tag{7.3.13}$$

(7.3.12)–(7.3.13) with equality sign again represent the system of hyperbolas whose properties are quite similar to the one for the case $u_0 > 0$.

For completeness, we provide a snapshot of the field plot at time $t = 0$ and $t = 0.5$ for values of parameters in *Regions 1, 2* and *3*, in Figs. 7.10, 7.11 and 7.12 respectively. Figure 7.13 gives the field plot at time $t = 0$ and $t = 0.5$ corresponding to a bifurcation point on the boundary between *Regions 1* and *2*. Similarly, Fig. 7.14 provides the field plot at time $t = 0$ and $t = 0.5$ for a bifurcation point on the boundary between *Regions 2* and *3*.

7.4 Dynamics of Evolution of Forward Sensitivities

Consider the tracer dynamics

$$\dot{\mathbf{x}} = \mathbf{f}(\mathbf{x}, \overline{\alpha}) + \mathbf{g}(\mathbf{x}, \overline{\alpha}) \tag{7.4.1}$$

where $\overline{\alpha} = (u_0, \alpha^T) \in \mathbb{R}^4$, $\mathbf{f} = (f_1, f_2)^T$ given in (7.1.15), $\mathbf{g} = (g_1, g_2)^T$ given in (7.1.16) and $u_1(t)$, $v_1(t)$ and $h_1(t)$ are given in (7.1.13). Substituting (7.1.15), (7.1.16) and (7.1.13) into (7.4.1), the latter becomes

$$\dot{x} = u(t) = F_1(\mathbf{x}, \overline{\alpha}) = -2\pi L \sin(2\pi K x) \cos(2\pi L y) u_0$$

$$+ \cos(2\pi M y) \left[\frac{\left(\cos\left(t\sqrt{4\pi^2 M^2 + 1}\right) + 4M^2 \pi^2 \right)}{4\pi^2 M^2 + 1} u_1(0) \right.$$

$$+ \frac{\sin\left(t\sqrt{4\pi^2 M^2 + 1}\right)}{\sqrt{4\pi^2 M^2 + 1}} v_1(0) + \frac{2\pi M \left(\cos\left(t\sqrt{4\pi^2 M^2 + 1}\right) - 1 \right)}{4\pi^2 M^2 + 1} h_1(0) \left. \right],$$

$$\dot{y} = v(t) = F_2(\mathbf{x}, \overline{\alpha}) = 2\pi K \cos(2\pi K x) \sin(2\pi L, y) u_0$$

$$- \cos(2\pi M y) \left[\frac{\sin\left(t\sqrt{4\pi^2 M^2 + 1}\right)}{\sqrt{4\pi^2 M^2 + 1}} u_1(0) \right.$$

$$- \cos\left(t\sqrt{4\pi^2 M^2 + 1}\right) v_1(0) + \frac{2\pi M \sin\left(t\sqrt{4\pi^2 M^2 + 1}\right)}{\sqrt{4\pi^2 M^2 + 1}} h_1(0) \left. \right],$$

$$\tag{7.4.2}$$

Fig. 7.10 A snapshot of the
time varying vector field
given by (7.1.14) at time
$t = 0$ and $t = 0.5$, where
$u_0 = 1.0$,
$u_1(0) = 12.5664$, $v_1(0) = \lambda$
and $h_1(0) = -2.0$. This
corresponds to
$V(0) = 1$, $U(0) = 4\pi$ and
$H(0) = 0$, which is a point in
Region 1 in Fig. 7.9. (**a**) Time
$t = 0.0$. (**b**) Time $t = 0.5$

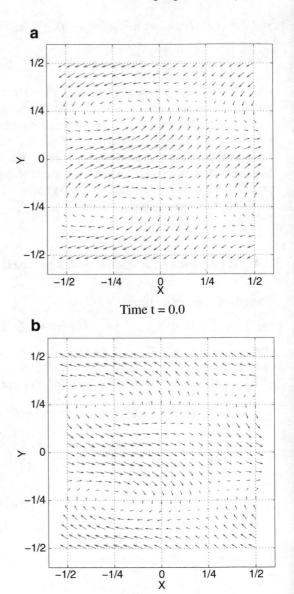

Time t = 0.0

Time t = 0.5

Fig. 7.11 A snapshot of the time varying vector field given by (7.1.14) $t = 0$ and $t = 0.5$, where $u_0 = 1.0$, $u_1(0) = 0.0$, $v_1(0) = \lambda$ and $h_1(0) = 0.0$ This corresponds to $V(0) = 1$, $U(0) = 0$ and $H(0) = 0$, which is a point in *Region 2* in Fig. 7.9. (**a**) Time $t = 0.0$. (**b**) Time $t = 0.5$

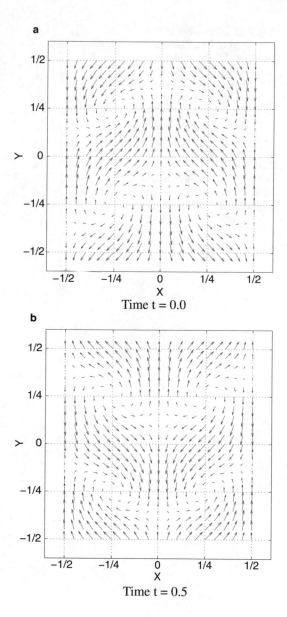

Fig. 7.12 A snapshot of the time varying vector field given by (7.1.14) A snapshot of the time varying vector field given by (3.31) at time $t = 0$ and $t = 0.5$, where $u_0 = 1.0$, $u_1(0) = -12.5664$, $v_1(0) = \lambda$ and $h_1(0) = 2.0$. This corresponds to $V(0) = 1$, $U(0) = -4\pi$ and $H(0) = 0$, which is a point in *Region 3* in Fig. 7.9. (**a**) Time $t = 0.0$. (**b**) Time $t = 0.5$

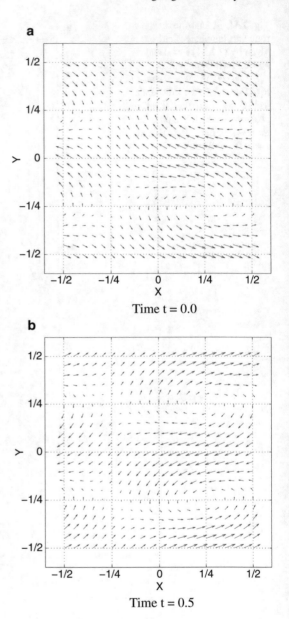

Fig. 7.13 A snapshot of the time varying vector field given by (7.1.14) A snapshot of the time varying vector field given by (3.31) at time $t = 0$ and $t = 0.5$, where $u_0 = 1.0$, $u_1(0) = 7.2832, v_1(0) = \lambda$ and $h_1(0) = -1.1592$. This corresponds to $V(0) = 1$, $U(0) = 2\pi + 1$ and $H(0) = 0$, which is a bifurcation point on the hyperbola separating *Regions 1* and *2* in Fig. 7.9. (**a**) Time $t = 0.0$. (**b**) Time $t = 0.5$

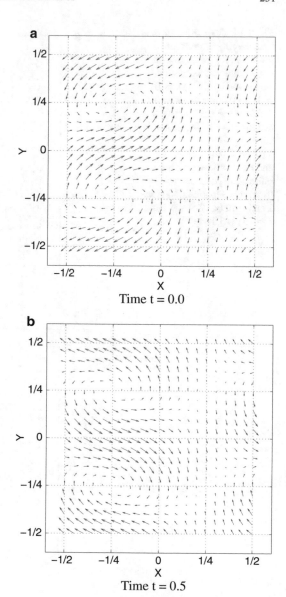

Fig. 7.14 A snapshot of the
time varying vector field
given by (7.1.14) A snapshot
of the time varying vector
field given by (3.31) $t = 0$
and $t = 0.5$, where $u_0 = 1.0$,
$u_1(0) = -7.2832, v_1(0) = \lambda$
and $h_1(0) = 1.1592$. This
corresponds to
$V(0) = 1, U(0) = -2\pi - 1$
and $H(0) = 0$, which is a
bifurcation point on the
hyperbola separating *Regions
2* and *3* in Fig. 7.9. (**a**) Time
$t = 0.0$. (**b**) Time $t = 0.5$

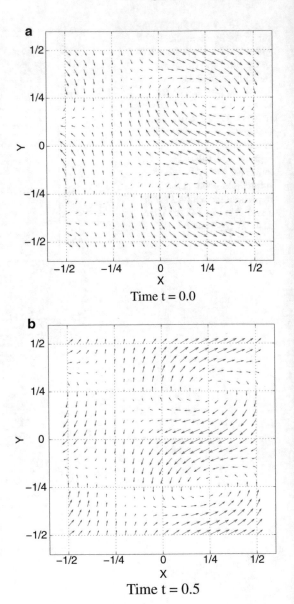

Now rewriting (7.4.1)–(7.4.2) succinctly as

$$\dot{\mathbf{x}} = \mathbf{F}(\mathbf{x}, \overline{\alpha}) \tag{7.4.3}$$

where $\mathbf{x} = (x, y)^T$, $\mathbf{F} : \mathbb{R}^2 \times \mathbb{R}^4 \to \mathbb{R}^2$, $\mathbf{F} = (F_1, F_2)^T$ defined in (7.4.2).
Discretizing (7.4.3) using the Euler scheme, we get

$$\mathbf{x}(k + 1) = \mathbf{x}(k) + \Delta t \mathbf{F}(\mathbf{x}(k), \overline{\alpha}), \tag{7.4.4}$$

where $t = k\Delta t$. Rewriting (7.4.4) in the standard form as

$$\mathbf{x}(k + 1) = \mathbf{M}(\mathbf{x}(k), \overline{\alpha}) \tag{7.4.5}$$

where $\mathbf{M} : \mathbb{R}^2 \times \mathbb{R}^4 \to \mathbb{R}^2$, $\mathbf{M} = (M_1, M_2)^T$ and

$$\mathbf{M}(\mathbf{x}, \overline{\alpha}) = \mathbf{x} + \Delta t F(\mathbf{x}, \overline{\alpha}). \tag{7.4.6}$$

Explicit form of M_1 and M_2 are given by

$$
\begin{aligned}
M_1\,(\mathbf{x}(k), \overline{\alpha}) &= x(k+1) \\
&= x(k) + \Big[-2\,\pi\,L\,\sin(2\,\pi\,K\,x(k))\,\cos(2\,\pi\,L\,y(t))\,u_0 \\
&\quad + \cos(2\,\pi\,M\,y(k))\left(\frac{\left(\cos\!\left(t\,\sqrt{4\,\pi^2\,M^2 + 1}\right) + 4M^2\,\pi^2\right)}{4\,\pi^2\,M^2 + 1}u_1(0)\right.\\
&\quad + \frac{\sin\!\left(t\,\sqrt{4\,\pi^2\,M^2 + 1}\right)}{\sqrt{4\,\pi^2\,M^2 + 1}}v_1(0) \\
&\quad \left.+ \frac{2\,\pi\,M\left(\cos\!\left(t\,\sqrt{4\,\pi^2\,M^2 + 1}\right) - 1\right)}{4\,\pi^2\,M^2 + 1}h_1(0)\right)\Big]\,\Delta t, \tag{7.4.7}
\end{aligned}
$$

$$
\begin{aligned}
M_2\,(\mathbf{x}(k), \alpha) &= y(k+1) \\
&= y(k) + \Big[2\,\pi\,K\,\cos(2\,\pi\,K\,x(k))\,\sin(2\,\pi\,L\,y(t))\,u_0 \\
&\quad - \cos(2\,\pi\,M\,y(k))\left(\frac{\sin\!\left(t\,\sqrt{4\,\pi^2\,M^2 + 1}\right)}{\sqrt{4\,\pi^2\,M^2 + 1}}u_1(0)\right.\\
&\quad - \cos\!\left(t\,\sqrt{4\,\pi^2\,M^2 + 1}\right)v_1(0) \\
&\quad \left.+ \frac{2\,\pi\,M\,\sin\!\left(t\,\sqrt{4\,\pi^2\,M^2 + 1}\right)}{\sqrt{4\,\pi^2\,M^2 + 1}}h_1(0)\right)\Big]\,\Delta t. \tag{7.4.8}
\end{aligned}
$$

with $t = k\Delta t$.

The elements of the Jacobian of M with respect to $\mathbf{x}(0)$

$$\mathbf{D}_{\mathbf{x}(0)}(\mathbf{M}) = \begin{bmatrix} \dfrac{\partial M_1}{\partial x(0)} & \dfrac{\partial M_1}{\partial y(0)} \\[2ex] \dfrac{\partial M_2}{\partial x(0)} & \dfrac{\partial M_2}{\partial y(0)} \end{bmatrix}$$

are given by

$$\mathbf{D}_{\mathbf{x}(0)}(\mathbf{M})_{(1,1)} = 1 + 4 K L \pi^2 u_0 \cos(2 \pi K x) \cos(2 \pi L y), \tag{7.4.9}$$

$$\mathbf{D}_{\mathbf{x}(0)}(\mathbf{M})_{(1,2)} = 4 l^2 \pi^2 u_0 \sin(2 \pi K x) \sin(2 \pi L y) \tag{7.4.10}$$

$$- 2 \pi M \sin(2 \pi M y) \left(\frac{v_1(0) \sin\left(t \sqrt{4 \pi^2 M^2 + 1}\right)}{\sqrt{4 \pi^2 M^2 + 1}} \right.$$

$$+ \frac{u_1(0) \left(\cos\left(t \sqrt{4 \pi^2 M^2 + 1}\right) + 4 M^2 \pi^2 \right)}{4 \pi^2 M^2 + 1}$$

$$\left. + \frac{2 \pi M h_1(0) \left(\cos\left(t \sqrt{4 \pi^2 M^2 + 1}\right) - 1 \right)}{4 \pi^2 M^2 + 1} \right),$$

$$\mathbf{D}_{\mathbf{x}(0)}(\mathbf{M})_{(2,1)} = -4 K^2 \pi^2 u_0 \sin(2 \pi K x) \sin(2 \pi L y), \quad \text{and} \tag{7.4.11}$$

$$\mathbf{D}_{\mathbf{x}(0)}(\mathbf{M})_{(2,2)} = 1 + 4 K L \pi^2 u_0 \cos(2 \pi K x) \cos(2 \pi L y)$$

$$+ 2 \pi M \sin(2 \pi M y) \left(\frac{u_1(0) \sin\left(t \sqrt{4 \pi^2 M^2 + 1}\right)}{\sqrt{4 \pi^2 M^2 + 1}} \right.$$

$$- v_1(0) \cos\left(t \sqrt{4 \pi^2 M^2 + 1}\right)$$

$$\left. + \frac{2 \pi M \sin\left(t \sqrt{4 \pi^2 M^2 + 1}\right)}{\sqrt{4 \pi^2 M^2 + 1}} h_1(0) \right) \tag{7.4.12}$$

Similarly, the elements of the Jacobian

$$\mathbf{D}_{\bar{\alpha}}(\mathbf{M}) = \begin{bmatrix} \dfrac{\partial M_1}{\partial u_0} & \dfrac{\partial M_1}{\partial u_1(0)} & \dfrac{\partial M_1}{\partial v_1(0)} & \dfrac{\partial M_1}{\partial h_1(0)} \\[2ex] \dfrac{\partial M_2}{\partial u_0} & \dfrac{\partial M_2}{\partial u_1(0)} & \dfrac{\partial M_2}{\partial v_1(0)} & \dfrac{\partial M_2}{\partial h_1(0)} \end{bmatrix}$$

are given by

$$\mathbf{D}_{\overline{\alpha}}(\mathbf{M})_{(1,1)} = -2\pi L \cos(2\pi L y) \sin(2\pi K x), \tag{7.4.13}$$

$$\mathbf{D}_{\overline{\alpha}}(\mathbf{M})_{(1,2)} = \frac{\cos(2\pi M y)\left(\cos\left(t\sqrt{4\pi^2 M^2 + 1}\right) + 4M^2\pi^2\right)}{4\pi^2 M^2 + 1}, \tag{7.4.14}$$

$$\mathbf{D}_{\overline{\alpha}}(\mathbf{M})_{(1,3)} = \frac{\sin\left(t\sqrt{4\pi^2 M^2 + 1}\right)\cos(2\pi M y)}{\sqrt{4\pi^2 M^2 + 1}}, \tag{7.4.15}$$

$$\mathbf{D}_{\overline{\alpha}}(\mathbf{M})_{(1,4)} = \frac{2\pi M \cos(2\pi M y)\left(\cos\left(t\sqrt{4\pi^2 M^2 + 1}\right) - 1\right)}{4\pi^2 M^2 + 1}, \tag{7.4.16}$$

$$\mathbf{D}_{\overline{\alpha}}(\mathbf{M})_{(2,1)} = 2\pi K \cos(2\pi K x) \sin(2\pi L y), \tag{7.4.17}$$

$$\mathbf{D}_{\overline{\alpha}}(\mathbf{M})_{(2,2)} = -\frac{\sin\left(t\sqrt{4\pi^2 M^2 + 1}\right)\cos(2\pi M y)}{\sqrt{4\pi^2 M^2 + 1}}, \tag{7.4.18}$$

$$\mathbf{D}_{\overline{\alpha}}(\mathbf{M})_{(2,3)} = \cos\left(t\sqrt{4\pi^2 M^2 + 1}\right)\cos(2\pi M y), \tag{7.4.19}$$

$$\mathbf{D}_{\overline{\alpha}}(\mathbf{M})_{(2,4)} = -\frac{2\pi M \sin\left(t\sqrt{4\pi^2 M^2 + 1}\right)\cos(2\pi M y)}{\sqrt{4\pi^2 M^2 + 1}}. \tag{7.4.20}$$

Recall that

$$u_1(k) = \left[\frac{\partial x(k)}{\partial x(0)}\right] = \begin{bmatrix} \dfrac{\partial x(k)}{\partial x(0)} & \dfrac{\partial x(k)}{\partial y(0)} \\[2mm] \dfrac{\partial y(k)}{\partial x(0)} & \dfrac{\partial y(k)}{\partial y(0)} \end{bmatrix}$$

is the forward sensitivity matrix of $x(k)$ with respect to the initial condition $x(0)$ and

$$V_1(k) = \left[\frac{\partial x(k)}{\partial \overline{\alpha}}\right] = \begin{bmatrix} \dfrac{\partial x(k)}{\partial u_0} & \dfrac{\partial x(k)}{\partial u_1(0)} & \dfrac{\partial x(k)}{\partial v_1(0)} & \dfrac{\partial x(k)}{\partial h_1(0)} \\[2mm] \dfrac{\partial y(k)}{\partial u_0} & \dfrac{\partial y(k)}{\partial u_1(0)} & \dfrac{\partial y(k)}{\partial v_1(0)} & \dfrac{\partial y(k)}{\partial h_1(0)} \end{bmatrix}$$

is the forward sensitivity matrix of $x(k)$ with respect to the parameters $\overline{\alpha} = (u_0, u_1(0), v_1(0), h_1(0))^T$.

From Chap. 4 the dynamics of evolution of $u_1(k)$ and $v_1(k)$ are given by

$$u_1(k+1) = \mathbf{D}_{x(0)}(\mathbf{M})u_1(k), \quad \text{with } u_1(0) = \mathbf{I} \tag{7.4.21}$$

and

$$v_1(k+1) = \mathbf{D}_{x(0)}(\mathbf{M})v_1(k) + \mathbf{D}_{\overline{\alpha}}(\mathbf{M}), \quad \text{with } v_1(0) = 0. \tag{7.4.22}$$

Table 7.1 Base values of
initial conditions $\mathbf{x}(0)$ and
control vector $\overline{\alpha}$

u_0	$u_1(0)$	$v_1(0)$	$h_1(0)$	x_0	y_0	t	Δt
1.000	0.000	0.500	0.000	0.0	0.0	0.5	2.5e−05

7.5 Sensitivity Plots

Recall that there are two initial conditions $(x_0, y_0)^T \in \mathbb{R}^2$ and four parameters $\overline{\alpha} = (u_0, u_1(0), v_1(0), h_1(0))^T \in \mathbb{R}^4$. From this infinite set of possibilities, in this section, for the lack of space, we present one typical example of the evolution of forward sensitivities starting from a base values for initial conditions and for parameters in Table 7.1.

7.5.1 Sensitivity to Initial Conditions

Evolution of the sensitivities of $\mathbf{x}(k)$ with respect to x_0 and y_0 are given in Figs. 7.15, 7.16, 7.17, 7.18. From Figs. 7.15, 7.16 it follows that sensitivity of $x(k)$ with respect to x_0 and y_0 are mirror images of each other. But the sensitivity of $y(k)$ with respect to x_0 as seen from Fig. 7.17 is very low for quite some time but becomes large positivity before decreasing to a low value. The sensitivity of $y(k)$ with respect to y_0 as seen from Fig. 7.18 remains low and positive for half of the interval but becomes large negative in time.

7.5.2 Sensitivity to Parameters

The evaluation of sensitivities of $x(k)$ and $y(k)$ with respect to $u_0, u_1(0), v_1(0)$ and $h_1(0)$ are given in Figs. 7.19, 7.20, 7.21, 7.22, 7.23, 7.24, 7.25, 7.26.

We invite the reader to systematically examine the sensitivity plots for different combination of initial conditions and parameters.

7.6 Data Assimilation Using FSM

In the previous section, we have analyzed sensitivity of tracer dynamics to the initial conditions $\mathbf{x}(0)$ and elements of the control vector $\overline{\alpha}$. These sensitivity plots help us judge the effectiveness of the distribution of observations in time.

In this section, we demonstrate the effectiveness of the forward sensitivity method (FSM) by conducting a series of numerical experiments involving data assimilation. We vary the number of observations, their temporal distribution and the observational error variance. To compare the effectiveness of data assimilation, we introduce the well known *root-mean-square error* (RMSE), given by

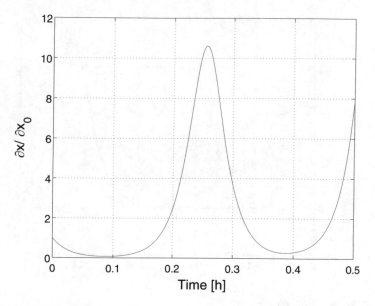

Fig. 7.15 Sensitivity of $x(t)$ w.r.t. $x(0)$

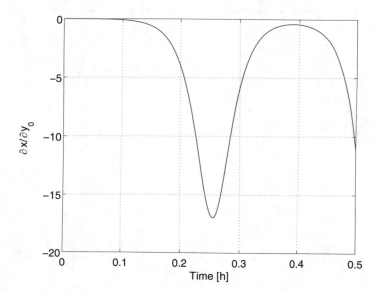

Fig. 7.16 Sensitivity of $x(t)$ w.r.t. $y(0)$

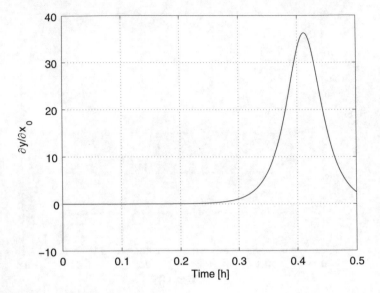

Fig. 7.17   Sensitivity of $y(t)$ w.r.t. $x(0)$

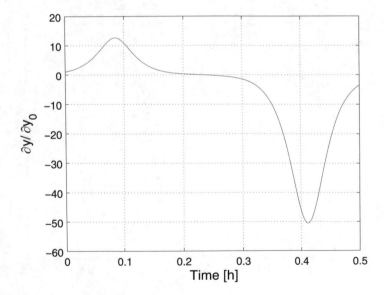

Fig. 7.18   Sensitivity of $y(t)$ w.r.t. $y(0)$

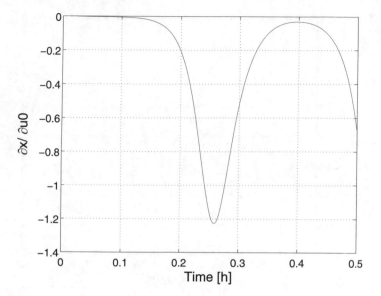

Fig. 7.19   Sensitivity of $x(t)$ w.r.t. u_0

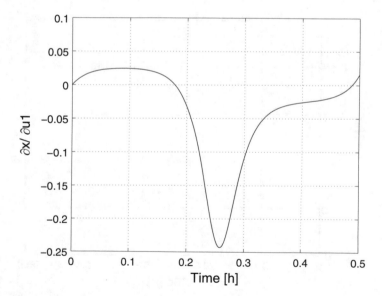

Fig. 7.20   Sensitivity of $x(t)$ w.r.t. $u_1(0)$

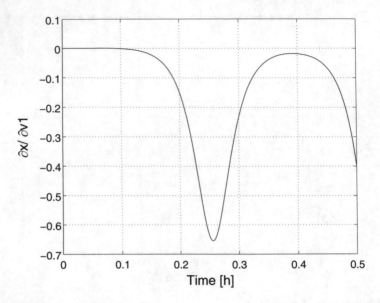

Fig. 7.21 Sensitivity of $x(t)$ w.r.t. $v_1(0)$

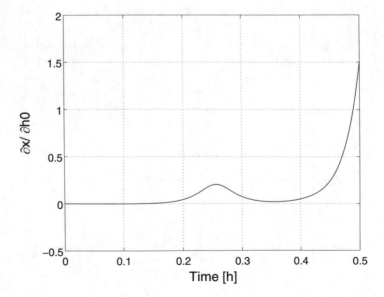

Fig. 7.22 Sensitivity of $x(t)$ w.r.t. $h_1(0)$

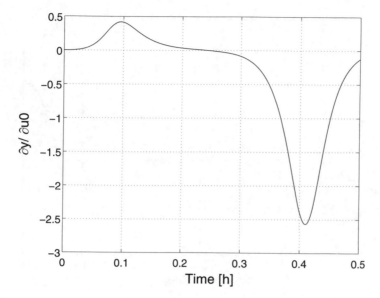

Fig. 7.23   Sensitivity of $y(t)$ w.r.t. u_0

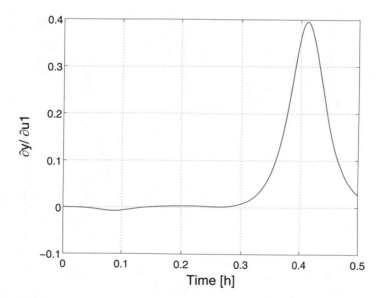

Fig. 7.24   Sensitivity of $y(t)$ w.r.t. $u_1(0)$

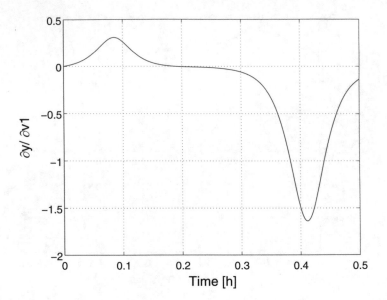

Fig. 7.25 Sensitivity of $y(t)$ w.r.t. $v_1(0)$

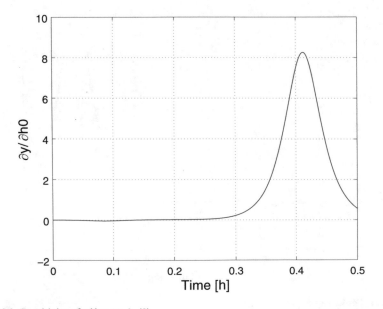

Fig. 7.26 Sensitivity of $y(t)$ w.r.t. $h_1(0)$

$$RMSE = \sqrt{\frac{\sum_{t=1}^{n}(\eta_t - \hat{\eta}_t)^2}{n}},$$ (7.6.1)

where η_t is the actual observation and $\hat{\eta}_t$ is the model prediction. It represents a very valuable measure of forecast errors and it allows for meaningful comparison different number of observations and their placement.

Another useful measure is the condition number of a matrix that arises in solving the least squares problem associated with FSM. The *condition number* of a matrix **A** is defined by Lewis et al. (2006, Appendix B).

$$\kappa(\mathbf{A}) = \|\mathbf{A}\| \|\mathbf{A}^{-1}\|.$$ (7.6.2)

where $\|\mathbf{A}\|$ is the norm. The condition number of a matrix **A** is a measure of the sensitivity of the solution of $\mathbf{Ax} = \mathbf{b}$ due to errors in the matrix **A** and the right hand side **b**. It can be said that matrices with small condition number are well conditioned; singular matrices have infinite condition numbers. We will use a *spectral condition number* defined on (Lewis et al. 2006).

$$\kappa_2(\mathbf{A}) = \|\mathbf{A}\|_2 \|\mathbf{A}^{-1}\|_2 = \frac{\max_i |\lambda_i|}{\min_i |\lambda_i|}.$$ (7.6.3)

7.6.1   Methodology

The observations that are used for data assimilation experiments are created using "base" trajectory by adding the noise with the observational error variance σ of different magnitudes given in Table 7.2. In every experiment, control vectors used for "base" trajectory are shown alongside the perturbed control vectors that we try to improve.

We start each numerical experiment with the incorrect control vector. The forward sensitivity method uses sensitivity of the forecast arising from this incorrect control. Our numerical experiments use observations from four different temporal

Table 7.2 Values of the observational error variance σ^2 used in data assimilation experiments to create perturbed observations for data assimilation

	σ^2
1	0.0050
2	0.0075
3	0.0100
4	0.0125
5	0.0150
6	0.0175
7	0.0200

distributions described as: START, MIDDLE, FINISH, UNIFORM. They are described in the next section. We keep track of the condition number, the ratio of the largest to the smallest of $\mathbf{H}^T\mathbf{H}$, since the inverse of $\mathbf{H}^T\mathbf{H}$ influences the optimal adjustments to the control vector (refer to Chap. 4 for details). We use the iterative process correcting elements of the control vector. After first iteration, we apply the correction and repeat the process, that is, we make another forecast using the corrected control vector. We have shown results of the firs three steps of the iteration. Clearly, FSM improves forecast at each step; we track improvements by following RMS error between observations and the forecast.

Each of our data assimilation experiments consist of four different temporal distributions of data called: START, MIDDLE, FINISH and UNIFORM. START puts all measurements in the first third of the temporal domain, MIDDLE puts all measurements in the second third or the temporal domain, FINISH puts all measurements in the last third of the temporal domain, and finally UNIFORM distributes measurements over the entire period of modeling.

7.6.2 Data Assimilation Experiment

We first run the model from the base values given in Table 7.1. The observations are created by adding noise as below

$$z_1(k) = x(k) + \eta_1(k)$$

$$z_2(k) = y(k) + \eta_2(k)$$

where $\eta_i(k) \sim N(0, \sigma^2)$ where σ^2 are given in Table 7.2.

The forecast is then generated from the perturbed $x(0)$ and $\overline{\alpha}$ in Table 7.3.

Results of the data assimilation are summarized in Table 7.4. The effect of the data distribution becomes apparent from this table. Plots of the sample trajectories resulting from the data assimilation are given in Figs. 7.27, 7.28, 7.29, 7.30.

Experiments reported in Sect. 7.6 indicate that even when the condition number κ is of order of 10^{20}, there can be an improvement made to the control vector, if the location of measurements is in the time period when the model has high sensitivity to the control vector, as shown in Table 7.4.

Jabrzemski (2014) dissertation contains results of data assimilation experiments for various combination of values for $\mathbf{x}(0)$ and $\overline{\alpha}$. We encourage the reader to explore these further.

Table 7.3 Perturbed initial conditions $\mathbf{x}(0)$ and $\overline{\alpha}$

u_0	$u_1(0)$	$v_1(0)$	$h_1(0)$	x_0	y_0	t	Δ time
1.129419	0.170145	0.549086	−0.040774	0.000	0.000	0.5	2.5e−05

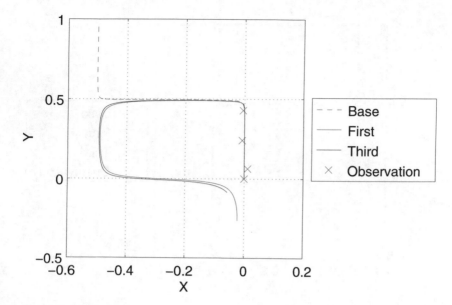

Fig. 7.27 Track START $\sigma^2 = 0.01$

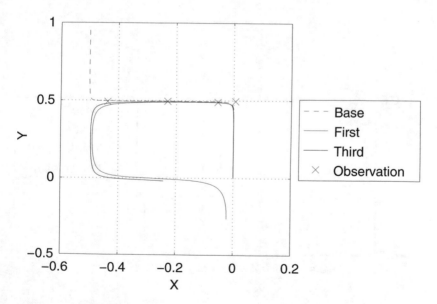

Fig. 7.28 Track MIDDLE $\sigma^2 = 0.01$

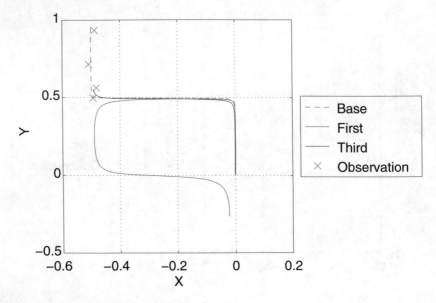

Fig. 7.29 Experiment 6.3: Track FINISH $\sigma^2 = 0.01$

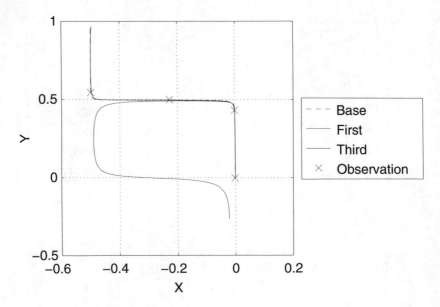

Fig. 7.30 Experiment 6.3: Track UNIFORM $\sigma^2 = 0.01$

Table 7.4 Comparison of data assimilations for a set of distributions and errors with four measurements, start (0,0)

Distribution scheme	Variance of observation noise	RMS Error		κ	
		Iteration		Iteration	
		First	Third	First	Third
START	0.005	0.0359	0.0275	5.3e+19	5.0e+19
	0.007	0.0691	0.0426	1.7e+20	1.4e+20
	0.010	0.0357	0.0281	2.5e+19	1.0e+20
	0.013	0.0169	0.0163	2.6e+20	1.1e+20
	0.015	0.0143	0.0143	1.4e+20	5.7e+19
	0.018	0.0469	0.0360	6.6e+20	3.7e+19
	0.020	0.0144	0.0143	8.3e+20	7.1e+20
MIDDLE	0.005	0.1249	0.0568	2.6e+21	2.7e+19
	0.007	0.1776	0.0482	1.3e+19	1.6e+19
	0.010	0.0890	0.0292	1.7e+18	3.5e+18
	0.013	0.0448	0.0164	9.4e+18	1.1e+19
	0.015	0.0367	0.0115	1.5e+19	3.9e+18
	0.018	0.0706	0.0347	5.8e+18	8.5e+19
	0.020	0.0593	0.0406	5.2e+19	2.2e+19
FINISH	0.005	0.7448	0.5471	2.9e+18	6.3e+17
	0.007	0.3430	0.0838	8.9e+18	1.1e+18
	0.010	0.7248	0.5323	9.6e+18	1.6e+18
	0.013	0.6104	0.4216	1.8e+20	9.1e+18
	0.015	0.4433	0.0225	1.8e+21	1.4e+19
	0.018	0.5804	0.3634	1.5e+18	4.0e+18
	0.020	0.5687	0.0187	1.0e+19	2.8e+19
UNIFORM	0.005	0.2856	0.1408	4.8e+20	1.4e+21
	0.007	0.1190	0.1053	2.6e+20	5.0e+19
	0.010	0.2664	0.0503	1.1e+20	7.9e+20
	0.013	0.1738	0.0402	7.9e+20	4.7e+22
	0.015	0.0559	0.0277	3.4e+21	4.5e+22
	0.018	0.1651	0.0862	1.3e+21	1.2e+20
	0.020	0.1081	0.0178	1.8e+21	2.3e+21

7.7   Notes and References

This chapter follows the developments in Jabrzemski and Lakshmivarahan (2013, 2014) and the dissertation by Jabrzemski (2014).

Appendix

Bounds on $u_1(t)$ in (7.1.13)

Let A and B be two-real–numbers—with at least one of them non-zero. For $0 \le x \le 2\pi$, consider the function

$$f(x) = A \cos x + B \sin x. \tag{7.A.1}$$

The behavior of f(x) in the interval $[0, 2\pi]$ is given in the following

Property 7.A.1. (a) $f(x)$ attains its extremum values at x^* where $\tan x^* = B/A$
(b) The distribution of $(x^*, f(x^*))$ pair for various values and signs of A and B are given in Table 7.5.
(c) If $f(x^*)$ is a maximum/minimum, then $f(x^* + \pi)$ is a minimum/maximum.
(d) For all $x \in [0, 2\pi]$,

$$-\sqrt{A^2 + B^2} = f_{min} \le f(x) \le f_{max} = \sqrt{A^2 + B^2}. \tag{7.A.2}$$

Proof. Setting the derivative of $f(x)$ to zero, the claim (a) follows. Hence

$$\cos x^* = \frac{A}{\sqrt{A^2 + B^2}} \quad \text{and} \quad \sin x^* = \frac{B}{\sqrt{A^2 + B^2}}. \tag{7.A.3}$$

Consider the case when $A > 0$ and $B > 0$ and $A > B$. For this choice of A and B, it can be verified that $0 \le x^* \le \pi/4$. Further, it is easy to verify that the second derivative of $f(x)$ is negative at x^* and positive at $(x^* + \pi)$. Hence $f(x^*)$ is a maximum and $f(x^* + \pi)$ is a minimum from which one of the entries in Table 7.5 and claim (c) follows. Considering these and (7.A.3) with (7.A.1), the claim (d) immediately follows

Table 7.5 Distribution of the pairs $(x^*, f(x^*))$

Ratio	A	B	x^* belongs to	$f(x) =$
$\|A\| > \|B\|$	+	+	$[0, \pi/4]$	f_{max}
or	−	−	$[0, \pi/4]$	f_{min}
$0 \le \frac{\|B\|}{\|A\|} \le 1$	−	+	$[-\pi/4, 0]$	f_{min}
	+	−	$[-\pi/4, 0]$	f_{max}
$\|A\| < \|B\|$	+	+	$[\pi/4, \pi/2]$	f_{max}
or	−	−	$[\pi/4, \pi/2]$	f_{min}
$\frac{\|B\|}{\|A\|} \ge 1$	−	+	$[-\pi/2, -\pi/4]$	f_{max}
	+	−	$[-\pi/2, -\pi/4]$	f_{min}

Definition and Properties of Standard Hyperbola

Equation for a standard hyperbola with center at (x_0, y_0) is given by (Spiegel 1968)

$$\frac{(x - x_0)^2}{a^2} - \frac{(y - y_0)^2}{b^2} = 1 \qquad (7.A.4)$$

where a and b are called the length of the semi major and semi minor axes respectively. The graph of the hyperbola in (7.A.4) is given in Fig. 7.31.

Fig. 7.31 An illustration of the standard hyperbola given by (7.A.4).
$AA' = 2a, BB' = 2b, c$ is the center, F and F' are foci.
$CF = CF' = \sqrt{a^2 + b^2}$. The eccentricity $c = \frac{\sqrt{a^2+b^2}}{a} > 1$

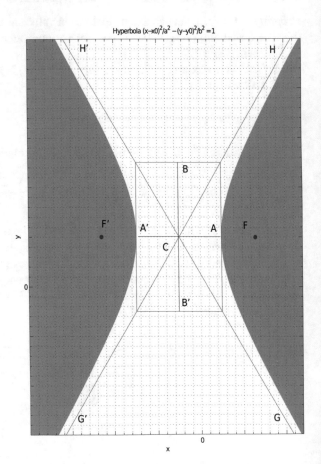

Epilogue

Computational power has now made it possible to generate $\mathcal{O}(100)$—of the order of 100, ensemble forecasts every 12 h at operational weather prediction centers around the world. Impressive yes, especially in the presence of the current dynamical models that require a billion—10^9—elements of control (initial conditions, boundary conditions, parameters). Yet, controversy surrounds this accomplishment. In short, a critical question arises: Is it more important to concentrate computer power on increasing the number of members in the ensemble, or should we devote more of the computational power to deterministic data assimilation that is designed to faithfully integrate observations into the dynamical constraints—*fit models to data.* Some support for this latter emphasis comes from noted chaos theorist Edward Lorenz. In his last recorded interview (Reeves 2014), he was asked to suggest research thrusts that had likelihood of improving weather forecasts. Your first thought might have been his advocacy for increasing the number of members in the ensemble, or improving the parameterization of physical processes such as change of phase of water processes, but no, he advocated improvement in data assimilation processes. His quotation from the interview follows:

> The first thing I would say is that current numerical prediction output[s] are much better than I ever thought they would be at this time. I wasn't sure they would ever get as good as they are now-certainly not within my lifetime... [and] makes me hopeful that we may actually get good forecasts a couple of weeks ahead some time. I still don't hold much hope for day-to-day forecasting a month ahead... I gather that a lot of the improvements have come from the improvement in initial conditions, and in turn, improvement in data assimilation methods... And I'm convinced that there will someday be better methods of data assimilation which incorporate the nonlinearity better than we are able to now in the assimilation process.

And by use of the word "nonlinearity", what did he mean? He meant the development of methodologies that went beyond the first-order Taylor series expansion of terms in the key functionals—the sum of squared differences between forecasts and observations. And we suspect that he was also advocating the inclusion of physical insight into the data assimilation process—what meteorological fields have the greatest impact on forecasts of the phenomenon, and what is the ideal placement and timing of observations that define the structure of these fields? Truly, the effort needs to combine input from the "observationalist"—that scientist who concentrates on understanding the processes that define the phenomenon—and those who concentrate on the formulation of the problem from a mathematical viewpoint—coupling observations, dynamics, and optimization strategies.

The Forecast Sensitivity Method (FSM) of dynamic data assimilation delivers information on the best places to take observation in order to correct elements of control. It is costly from a computational viewpoint since it requires evolution of sensitivity as input to the least squares minimization process. The number of sensitivity equations is equal to the number of control elements—an immense number for models such as the operational numerical weather prediction (NWP) models. And this number is associated with a specific model output—another output requires the same number of equations. The nonlinearity that Lorenz mentions is handled by this process, yet insight into the processes that govern the phenomenon can also be used to advantage.

To see how the physical insight comes into play, consider the following example related to severe convective storm prediction in mid-latitudes—thunderstorms and associated tornadoes in some instances. As Lorenz stated in the quotation above, and as seasoned weather forecasters know, the current NWP models give excellent guidance over a forecast horizon of several days in most weather regimes. In particular, the Storm Prediction Center of the National Weather Service uses operational deterministic weather prediction models to make 2-day predictions of storm-warning "boxes"—regions where severe storms are expected. These boxes typically cover areas of 100 miles by 100 miles (10^4 square miles). Restructuring the dimensions and areas covered by these boxes, of course, takes place in response to forecasts at later times. From geographical points within these boxes, 2-day back trajectories of air motion can be constructed from the model output. This helps define a subset of model region where sensitivity is expected to be great. This limits the number of control elements that need to be considered and can lead to more efficient (less time consuming) data assimilation. And in this age of mobile observation platforms and satellite observations, instruments can be moved to locations deemed most important for the assimilation process, possibly 6–12 h ahead of expected time for storm outbreak.

We should also mention the value of Pontryagin Minimum Principle (PMP) as a method of reducing forecast error in these practical problems. In certain weather regimes such as the Gulf Return Flow as discussed in Chap. 6, bias is present in the operational forecasts—typically in moisture content of air returning to the continent after modification over the Gulf. Consequences of incorrect moisture forecast in these situations can be critical as mentioned earlier. There are typically 25–30 such events every year between November and March. This group of events could be further subdivided based on source region of the air that enters the Gulf—Arctic outbreaks of cold air, maritime air masses that descend on the Gulf from the Pacific Northwest. Corrections to forecasts based on the PMP would give corrective terms to the dynamical constraints that would likely reduce the bias and improve the forecasts. Although this process does not easily identify the physical basis for the bias, the statistical corrections force a good fit between observations and model output.

As we look to the future of deterministic forecasting with dynamical models that are becoming ever-more complicated in response to the availability of increased computational power, the number of control elements invariably increases. Does this scenario translate into more accurate predictions? A good question and the answer is: "not necessarily." For example, imbalances in the governing equations, physically-, economically-, or biologically-based imbalances dependent on the phenomenon under investigation, drive changes that often lead to exponential growth of the disturbances (in weather: the incipient cyclone, storm clouds, in medicine: heart rate, glucose concentration). The natural-process growth also leads to exponential growth of the forecast error and limits the predictability. Thus, it becomes important to adjust control based on judiciously placed observations in association with least squares fit. The FSM holds promise for helping solve this challenging problem.

The task that lies before us points to methodologies that account for nonlinearity. We have shown how second-order sensitivity—account for nonlinearity in the assimilation process—can be incorporated into FSM. It can also be incorporated into 4D-VAR (See Chap. 25 in Lewis et al. 2006). However, the application of these ideas to our most complex models has not been tested. This is the vista yet unexplored. The challenge is not simply to take advantage of the ever-increasing computer power, but to innovatively explore these second-order methods with an eye on consequences of reducing the elements of control, strategic placement of observations, and unraveling the information contained in the evolution of sensitivities.

Appendix A
Basic Notation

In this Appendix we have assembled the notation and basic facts needed to pursue the development in the main text.

A.1 Basic Notation

- $\mathbb{R}$—the set of all real numbers
- $\mathbb{R}^n$—the set of all real column vectors of size n. if $\mathbf{x} \in \mathbb{R}^n$, then $\mathbf{x} = (x_1, x_2, \ldots, x_n)^T$, where T denotes the transpose operation.
- $\mathbb{R}^{n \times m}$—the set of all real $n \times m$ matrices. If $\mathbf{A} \in \mathbb{R}^{n \times m}$, then $\mathbf{A} = [a_{ij}]$ for $1 \leq i \leq n$ and $1 \leq j \leq m$. When $n = m$, $\mathbf{A}$ is a square matrix.

A.2 Inner Product

- If $\mathbf{a}, \mathbf{x} \in \mathbb{R}^n$, then $\langle \mathbf{a}, \mathbf{x} \rangle = \sum_{i=1}^{n} a_i x_i = \langle \mathbf{x}, \mathbf{a} \rangle$ is called the inner product.
- $\langle \mathbf{a}_1 + \mathbf{a}_2, \mathbf{x} \rangle = \langle \mathbf{a}_1, \mathbf{x} \rangle + \langle \mathbf{a}_2, \mathbf{x} \rangle$
- $\langle \alpha \mathbf{a}, \mathbf{x} \rangle = \alpha \langle \mathbf{a}, \mathbf{x} \rangle$; $\langle \mathbf{a}, \alpha \mathbf{x} \rangle = \alpha \langle \mathbf{a}, \mathbf{x} \rangle$ for any real α
- $\langle \mathbf{A}\mathbf{a}, \mathbf{x} \rangle = \langle \mathbf{a}, \mathbf{A}^T \mathbf{x} \rangle$ for $\mathbf{y} \in \mathbb{R}^n$ and $\mathbf{A} \in \mathbb{R}^{n \times n}$. This defines the adjoint $\mathbf{A}^T$ of $\mathbf{A}$.

A.3 Gradient, Jacobian, Hessian

- Let $f : \mathbb{R}^n \to \mathbb{R}$. *Gradient* $\nabla_x f$ of f with respect to x is a column vector of partial derivatives of f. Thus, $\nabla_x f = \left(\frac{\partial f}{\partial x_1}, \frac{\partial f}{\partial x_2}, \ldots, \frac{\partial f}{\partial x_n} \right)^T$.

© Springer International Publishing Switzerland 2017
S. Lakshmivarahan et al., *Forecast Error Correction using Dynamic Data Assimilation*, Springer Atmospheric Sciences, DOI 10.1007/978-3-319-39997-3

- *Hessian* $\nabla_x^2 f$ of $f : \mathbb{R}^n \to \mathbb{R}$ is a symmetric matrix of second partial derivatives of f. Thus

$$\nabla_x^2 f = \left[\frac{\partial^2 f}{\partial x_i \partial x_j} \right], 1 \leq i, j \leq n.$$

When $n = 2$,

$$\nabla_x^2 f = \begin{bmatrix} \frac{\partial^2 f}{\partial x_1^2} & \frac{\partial^2 f}{\partial x_1 \partial x_2} \\ \frac{\partial^2 f}{\partial x_2 \partial x_1} & \frac{\partial^2 f}{\partial x_2^2} \end{bmatrix},$$

which is symmetric since $\frac{\partial^2 f}{\partial x_1 \partial x_2} = \frac{\partial^2 f}{\partial x_2 \partial x_1}$

- Let $f : \mathbb{R}^n \to \mathbb{R}^m$, and $f(x) = (f_1(x), f_2(x), \ldots, f_n(x))^T$. Then , the *Jacobian* $\mathbf{D}_x(f)$ is an $n \times m$ matrix of partial derivatives of the component of f arranges in a matrix as follows:

$$\mathbf{D}_x(f) = \begin{bmatrix} \frac{\partial f_1}{\partial x_1} & \frac{\partial f_1}{\partial x_2} & \cdots & \frac{\partial f_1}{\partial x_n} \\ \frac{\partial f_2}{\partial x_1} & \frac{\partial f_2}{\partial x_2} & \cdots & \frac{\partial f_2}{\partial x_n} \\ & & \vdots & \\ \frac{\partial f_m}{\partial x_1} & \frac{\partial f_m}{\partial x_2} & \cdots & \frac{\partial f_m}{\partial x_n} \end{bmatrix} = \left[\frac{\partial f_i}{\partial x_j} \right], 1 \leq i \leq m, 1 \leq j \leq m$$

- Let $f(x) = \langle \mathbf{a}, \mathbf{x} \rangle = \sum_{i=1}^{n} a_i x_i$. Then $\nabla_x f(x) = \mathbf{a}$
- Let $f(x) = \mathbf{x}^T \mathbf{A} \mathbf{x}$, where $\mathbf{A}$ is a symmetric matrix.
  Then $\nabla_x f(x) = 2\mathbf{A}\mathbf{x}$.
- Let $f(x) = \langle \mathbf{a}, \mathbf{h}(x) \rangle = \sum_{i=1}^{n} a_i h_i(x)$, where $h_i : \mathbb{R}^n \to \mathbb{R}$. Then $\nabla_x f(x) = \sum_{i=1}^{n} a_i \nabla_x h_i(x) = \mathbf{D}_x^T(h) \mathbf{a}$. As an example, but $n = 2$ and $f(x) = a_1 h_1(x) + a_2 h_2(x)$. Then

$$\nabla_x f(x) = a_1 \begin{bmatrix} \frac{\partial h_1}{\partial x_1} \\ \frac{\partial h_1}{\partial x_2} \end{bmatrix} + a_2 \begin{bmatrix} \frac{\partial h_2}{\partial x_1} \\ \frac{\partial h_2}{\partial x_2} \end{bmatrix} = \begin{bmatrix} \frac{\partial h_1}{\partial x_1} & \frac{\partial h_2}{\partial x_1} \\ \frac{\partial h_1}{\partial x_2} & \frac{\partial h_2}{\partial x_2} \end{bmatrix} \begin{bmatrix} a_1 \\ a_2 \end{bmatrix} = \mathbf{D}_x^T(h) a$$

- Let $f(x) = \mathbf{h}^T(x) \mathbf{A} \mathbf{h}(x)$, $\mathbf{A}$—symmetric, then

$$\nabla f(x) = 2\mathbf{D}_x^T(h) \mathbf{A} \mathbf{h}(x).$$

As an example, let $n = 2$ and $\mathbf{A} = \begin{bmatrix} a_1 & b \\ b & a_2 \end{bmatrix}$.

Then

$$f(x) = (h_1, h_2) \begin{bmatrix} a_1 & b \\ b & a_2 \end{bmatrix} \begin{bmatrix} h_1 \\ h_2 \end{bmatrix} = a_1 h_1^2 + 2b h_1 h_2 + a_2 h_2^2$$

$$\nabla_x f(x) = 2 \begin{bmatrix} a_1 h_1 \frac{\partial h_1}{\partial x_1} + b h_1 \frac{\partial h_2}{\partial x_1} + b h_2 \frac{\partial h_1}{\partial x_1} + a_2 h_2 \frac{\partial h_1}{\partial x_1} \\ a_1 h_1 \frac{\partial h_2}{\partial x_2} + b h_1 \frac{\partial h_2}{\partial x_2} + b h_2 \frac{\partial h_1}{\partial x_2} + a_2 h_2 \frac{\partial h_2}{\partial x_2} \end{bmatrix}$$

$$= 2 \mathbf{D}_x^T(h) \mathbf{A} h(x)$$

A.4   Taylor Expansion

- Let $f : \mathbb{R} \to \mathbb{R}, x, y \in \mathbb{R}$ and y small.
  Then $f(x + y) = f(x) + \frac{df}{dx} y + \frac{1}{2} \frac{d^2 f}{dh^2} y^2 + \cdots$
- Let $f : \mathbb{R}^n \to \mathbb{R}, x, y \in \mathbb{R}^n$ and y small.
  Then $f(x + y) = f(x) + [\nabla_x f(x)]^T y + \frac{1}{2} y^T [\nabla_x^2 f(x)]^T y + \cdots$
- Let $f : \mathbb{R}^n \to \mathbb{R}^n, x, y \in \mathbb{R}^n$ and y small,
  and $f(x) = (f_1(x), f_2(x), \ldots, f_n(x))^T$.
  Then

$$f(x + y) = f(x) + \mathbf{D}_x(f) y + \frac{1}{2} \mathbf{D}_x^2(f, y) + \cdots,$$

where

$$D_x^2(f, y) = \left[y^T [\nabla_x^2 f_1(x)] y, y^T [\nabla_x^2 f_2(x)] y, \ldots, y^T [\nabla_x^2 f_n(x)] y \right]^T.$$

A.5   First Variations

- Let $f : \mathbb{R} \to \mathbb{R}$. The first variation δf of $f(x)$ induced by a change δx in x is given by

$$\delta f = \left(\frac{df}{dx} \right) \delta x.$$

- Let $f : \mathbb{R}^n \to \mathbb{R}$. Then

$$\delta f = \sum_{i=1}^{n} \left(\frac{\partial f}{\partial x_i} \right) \delta x_i = \langle \nabla_x f(x), \delta x \rangle$$

- Let $f(x) = \mathbf{a}^T\mathbf{x}$, then $\delta f = \langle a, \delta x \rangle$.
- Let $f(x) = \mathbf{a}^T\mathbf{h}(x)$, then $\delta f = \langle \mathbf{D}_x^T(h)a, \delta x \rangle$
- Let $f(x) = \mathbf{h}^T(x)\mathbf{A}\mathbf{h}(x)$, then $\delta f = \langle 2\mathbf{D}_x^T\mathbf{A}h(x), \delta x \rangle$

A.6   Conditions for Minima

- Let $f : \mathbb{R}^n \to \mathbb{R}$, which is twice differentiable.
- A necessary condition for the minimum of $f(x)$ at x^* is that the gradient must vanish, this is,

$$\nabla f(x)\Big|_{x=x^*} = 0$$

- A necessary and sufficient condition for the minimum of $f(x)$ at x^* is that the gradient vanishes and the Hessian is positive definite, that is $\nabla f(x)\Big|_{x=x^*} = 0$ and $\nabla^2 f(x)\Big|_{x=x^*}$ is positive definite.

References

Anthes RA (1974) Data assimilation and initialization of hurricane prediction models. J Atmos Sci 31(3):702–719

Apte A, Jones CK, Stuart A (2008) A Bayesian approach to Lagrangian data assimilation. Tellus A 60(2):336–347

Arakawa A (1966) Computational design for long-term numerical integration of the equations of fluid motion: two-dimensional incompressible flow. Part 1. J Comput Phys 1(1):119–143

Athans M, Falb PL (1966) Optimal control: an introduction to the theory and its applications. McGraw-Hill, New York

Ball F (1960) Control of inversion height by surface heating. Q J R Meteorol Soc 86(370):483–494

Bennett AF (1992) Inverse methods in physical oceanography. Cambridge University Press, Cambridge

Bennett S (1996) A brief history of automatic control. IEEE Control Syst 16(3):17–25

Bennett AF, Thorburn MA (1992) The generalized inverse of a nonlinear quasigeostrophic ocean circulation model. J Phys Oceanogr 22:213–230

Bode HW et al. (1945) Network analysis and feedback amplifier design. van Nostrand, New York

Boltânski VG (1978) Optimal control of discrete systems. Halsted Press, New York

Boltânski VG, Trirogoff K, Tarnove I, Leitmann G (1971) Mathematical methods of optimal control. Holt, Rinehart, and Winston, New York

Bryson AE (1975) Applied optimal control: optimization, estimation and control. CRC, Boca Raton

Bryson E Jr (1996) Optimal control-1950 to 1985. IEEE Control Syst 16(3):26–33

Bryson AE, Ho Y (1999) Dynamic optimization. Addison Wesley, Longman, Menlo Park

Burgers J (1939) Mathematical examples illustrating relations occuring in theory of turbulent fluid motion. Verh Kon Ned Akad Wet (Eerste Sectie) XVII(2):1–53

Cacuci DG (2003) Sensitivity and uncertainty analysis, volume I: theory, vol 1. CRC, Boca Raton

Cacuci DG, Ionescu-Bujor M, Navon IM (2005) Sensitivity and uncertainty analysis, volume II: applications to large-scale systems, vol 2. CRC, Boca Raton

Canon M (1970) Theory of optimal control and mathematical programming. McGraw-Hill, New York

Carrier G, Pearson C (1976) Partial differential equations: theory and technique. Academic, New York

Carter EF (1989) Assimilation of Lagrangian data into a numerical model. Dyn Atmos Oceans 13(3):335–348

Charney J (1966) Committee on atmospheric sciences. The feasibility of a global observation and analysis experiment. National Academy of Science-National Research Council (NAS-NRC) (published in abbreviated form in Bull Amer Meteor Soc) 47:200–220

Charney J (1969) Garp topics. Bull Amer Meteor Soc 50:136–141

Clebsch A (1857) Über eine allgemeine transformation der hydrodynamischen gleichungen. J Reine Angew Math 54:293–312

Crisp CA, Lewis JM (1992) Return flow in the Gulf of Mexico. Part i: A classificatory approach with a global historical perspective. J Appl Meteorol 31(8):868–881

Cruz JB (1982) System sensitivity analysis. Wiley, New York

Dee DP, Da Silva AM (1998) Data assimilation in the presence of forecast bias. Q J R Meteorol Soc 124(545):269–295

Deif A (2012) Sensitivity analysis in linear systems. Springer, Berlin

De Maria M, Jones RW (1993) Optimization of a hurricane track forecast model with the adjoint model equations. Mon. Weather Rev. 121:1730–1745

Derber JC (1989) A variational continuous assimilation technique. Mon Weather Rev 117(11):2437–2446

Eslami M (2013) Theory of sensitivity in dynamic systems: an introduction. Springer, Berlin

Fiacco A (1984) Sensitivity, stability, and parametric analysis. North-Holland, Amsterdam

Frank PM (1978) Introduction to system sensitivity theory. Academic, New York

Friedland B (1969) Treatment of bias in recursive filtering. IEEE Trans Autom Control 14(4):359–367

Friedman B (1956) Principles and techniques of applied mathematics. Wiley, New York

Gauss CF (1809) Theoria motus corporum coelestium in sectionibus conicis. Theory of the motion of heavenly bodies moving about the Sun in conic sections (English Trans by C. H. Davis reissued by Dover, 1963.)

Goldstein H (1950) Classical mechanics. Addison-Wesley world student series. Addison-Wesley, Reading, MA

Goldstein H (1980) A history of the calculus of variations from the 17th century through the 19th century, Springer Verlag, NY

Golub GH, Van Loan CF (2012) Matrix computations. Johns Hopkins University Press, Baltimore, MD

Griffith A, Nichols N (2000) Adjoint methods in data assimilation for estimating model error. Flow, turbulence and combustion 65(3–4):469–488

Gustafson TV, Kullenberg B (1936) Untersuchungen von Trägheitsströmungen in der Ostsee. 13, Svenska hydrografisk-biologiska

Hirsch MW, Smale S, Devaney RL (2004) Differential equations, dynamical systems, and an introduction to chaos. Academic, New York

Jabrzemski R (2014) Application of the forward sensitivity method to data assimilation of the Lagrangian tracer dynamics. PhD dissertation, School of Computer Science, University of Oklahoma, Norman, Oklahoma

Jabrzemski R, Lakshmivarahan S (2011) Five variable model. Technical report, School of Computer Science, University of Oklahoma

Jabrzemski R, Lakshmivarahan S (2013) Lagrangian data assimilation: part 1 analysis of bifurcations in tracer dynamics. In: 17th conference on integrated observing and assimilation systems for the atmosphere, oceans, and land surface. American Meteorological Society, Boston, MA

Jabrzemski R, Lakshmivarahan S (2014) Lagrangian data assimilation: part 2 data assimilation with the forward sensitivity method. In: 18th conference on integrated observing and assimilation systems for the atmosphere, oceans, and land surface. American Meteorological Society, Boston, MA

Keller HB (1976) Numerical solution of two point boundary value problems. SIAM, Philadelphia

Kokotovic P, Rutman R (1965) Sensitivity of automatic control systems-survey (automatic control system sensitivity and synthesis of self-adjusting systems). Autom Remote Control 26:727–749

Kuhn HW (1982) Nonlinear programming: a historical view. ACM SIGMAP Bull 31:6–18

Kuznetsov YA (2013) Elements of applied bifurcation theory. Springer, Berlin

Kuznetsov L, Ide K, Jones C (2003) A method for assimilation of Lagrangian data. Mon Weather Rev 131(10):2247–2260

Lakshmivarahan S, Dhall SK (1990) Analysis and design of parallel algorithms: arithmetic and matrix problems. McGraw-Hill, New York

Lakshmivarahan S, Lewis JM (2010) Forward sensitivity approach to dynamic data assimilation. Adv Meteorol 2010. doi:10.1155/2010/375615

Lakshmivarahan S, Lewis JM (2013) Nudging methods: a critical overview. In: Park S, Xu L (eds) Data assimilation for atmospheric, oceanic and hydrologic applications, vol II. Springer, Berlin, pp 27–57

Lakshmivarahan S, Honda Y, Lewis J (2003) Second-order approximation to the 3dvar cost function: application to analysis/forecast. Tellus A 55(5):371–384

Lakshmivarahan S, Lewis JM, Phan D (2013) Data assimilation as a problem in optimal tracking: application of Pontryagin's minimum principle. J Atmos Sci 70

Lewis JM (1972) An operational upper air analysis using the variational method. Tellus 24(6):514–530

Lewis JM (1993) Challenges and advantages of collecting upper-air data over the Gulf of Mexico. Mar Technol Soc J 27:56–65

Lewis JM (2005) Roots of ensemble forecasting. Mon Weather Rev 133(7):1865–1885

Lewis JM (2007) Use of a mixed-layer model to investigate problems in operational prediction of return flow. Mon Weather Rev 135(7):2610–2628

Lewis JM, Crisp CA (1992) Return flow in the Gulf of Mexico. Part ii: Variability in return-flow thermodynamics inferred from trajectories over the gulf. J Appl Meteorol 31(8):882–898

Lewis J, Lakshmivarahan S (2008) Sasaki's pivotal contribution: calculus of variations applied to weather map analysis. Mon Weather Rev 136(9):3553–3567

Lewis J, Lakshmivarahan S (2012) Sensitivity as a component of variational data assimilation. In: AGU fall meeting abstracts, vol 1, p 0277

Lewis JM, Lakshmivarahan S, Dhall S (2006) Dynamic data assimilation: a least squares approach. Cambridge University Press, Cambridge

Lewis FL, Syrmos VL (1995) Optimal control. Wiley, New York

Lilly DK (1968) Models of cloud-topped mixed layers under a strong inversion. Q J R Meteorol Soc 94(401):292–309

Lilly DK (1997) Computational design for long-term numerical integration of the equations of fluid motion: two-dimensional incompressible flow. Part 1. J Comput Phys 135:101–102

Liu Q, Lewis JM, Schneider JM (1992) A study of cold-air modification over the Gulf of Mexico using in situ data and mixed-layer modeling. J Appl Meteorol 31(8):909–924

Looper J (2013) Assessing impacts of precipitation and parameter uncertainty on distributed hydrologic modeling. PhD dissertation, School of Civil and Environmental Engineering, University of Oklahoma, Norman, Oklahoma

Lorenz EN (1963) Deterministic nonperiodic flow. J Atmos Sci 20(2):130–141

Lorenz E (1982) Atmospheric predictability experiments with a large numerical model. Tellus 34(6):505–513

Lorenz EN (1995) The essence of chaos. University of Washington Press, Seattle

Ménard R, Daley R (1996) The application of Kalman smoother theory to the estimation of 4dvar error statistics. Tellus A 48(2):221–237

Meyer CD (2000) Matrix analysis and applied linear algebra, vol 2. SIAM, Philadelphia

Molcard A, Piterbarg LI, Griffa A, Özgökmen TM, Mariano AJ (2003) Assimilation of drifter observations for the reconstruction of the Eulerian circulation field. J Geophys Res Oceans 108(C3)

Naidu DS (2002) Optimal control systems. CRC, Boca Raton

Nago NT (1971) Sensitivity of automatic control systems. Avtomat i Telemekh (5):53–82

Özgökmen TM, Molcard A, Chin TM, Piterbarg LI, Griffa A (2003) Assimilation of drifter observations in primitive equation models of midlatitude ocean circulation. J Geophys Res Oceans 108(C7)

Pant R (2012) Robust decision-making and dynamic resilience estimation for interdependent risk analysis. PhD dissertation. School of Industrial and Systems Engineering, University of Oklahoma, Norman, Oklahoma

Peitgen HO, Jürgens H, Saupe D (1992) Chaos and fractals. Springer, Berlin

Phan DT (2011) Data assimilation in the ecosystem carbon pool model using the forward sensitivity method. Master's thesis. School of Computer Science, University of Oklahoma, Norman, Oklahoma

Phillips O (1959) The scattering of gravity waves by turbulence. J Fluid Mech 5(2):177–192

Platzman GW (1964) An exact integral of complete spectral equations for unsteady one-dimensional flow1. Tellus 16(4):422–431

Polak E (2012) Optimization: algorithms and consistent approximations. Springer, Berlin

Pontryagin LS, Boltânski VG, Gamkrelidze RV, Mischenko EF (1962) The mathematical theory of optimal control processes. Interscience, New York

Rabitz H, Kramer M, Dacol D (1983) Sensitivity analysis in chemical kinetics. Annual review of physical chemistry 34(1):419–461

Rauch HE, Striebel C, Tung F (1965) Maximum likelihood estimates of linear dynamic systems. AIAA J 3(8):1445–1450

Reeves RW (2014) Edward lorenz revisiting the limits of predictability and their implications: an interview from 2007. Bull Am Meteorol Soc 95(5):681–687

Richtmyer RD, Morton KW (1967) Difference methods for initial-value problems. Interscience, Wiley, New York.

Roberts SM, Shipman JS (1972) Two-point boundary value problems: shooting methods, vol 31. Elsevier, Amsterdam

Ronen Y (1988) Uncertainty analysis. Franklin Book Company, Elkins Park, PA

Rozenwasser E, Yusupov R (1999) Sensitivity of automatic control systems. CRC, Boca Raton

Salman H, Ide K, Jones CK (2008) Using flow geometry for drifter deployment in Lagrangian data assimilation. Tellus A 60(2):321–335

Saltelli A, Chan K, Scott E (2000) Sensitivity analysis. Wiley, New York

Saltelli A, Tarantola S, Campolongo F, Ratto M (2004) Sensitivity analysis in practice: a guide to assessing scientific models. Wiley, New York

Saltelli A, Ratto M, Andres T, Campolongo F, Cariboni J, Gatelli D, Saisana M, Tarantola S (2008) Global sensitivity analysis: the primer. Wiley, New York

Saltzman B (1962) Finite amplitude free convection as an initial value problem-1. J Atmos Sci 19(4):329–341

Sasaki YK (1958) An objective analysis based on the variational method, vol 36. Journal of the Meteorological Society of Japan, Japan

Sasaki Y (1970a) Numerical variational analysis formulated under the constraints as determined by longwave equations and a low-pass filter. Mon Weather Rev 98(12):884–898

Sasaki Y (1970b) Numerical variational analysis with weak constraint and application to surface analysis of severe storm gust. Mon Weather Rev 98(12):899–910

Sasaki Y (1970c) Some basic formalisms in numerical variational analysis. Mon Weather Rev 98(12):875–883

Simmons A, Hollingsworth A (2002) Some aspects of the improvement in skill of numerical weather prediction. Q J R Meteorol Soc 128(580):647–677

Sobral M (1968) Sensitivity in optimal control systems. In: IEEE proceedings

Spiegel MR (1968) Mathematical handbook of formulas and tables. McGraw-Hill, New York

Sverdrup HU, Johnson MW, Fleming RH et al. (1942) The oceans: their physics, chemistry, and general biology. Prentice-Hall, New York

Symon KR (1971) Mechanics, 3rd edn. Addison-Wesley, New York

Tennekes H, Driedonks A (1981) Basic entrainment equations for the atmospheric boundary layer. Bound-Layer Meteorol 20(4):515–531

Tomović R, Vukobratovíc M (1972) General sensitivity theory. Elsevier, Amsterdam

Tromble E (2011) Parameter estimation for and aspects of coupled flood inundation using the ADCIRC hydrodynamic model. PhD dissertation, School of Civil and Environmental Engineering, University of Oklahoma, Norman, Oklahoma

Wiener N (1961) Cybernetics or control and communication in the animal and the machine. MIT, Cambridge, MA

Zupanski D (1997) A general weak constraint applicable to operational 4dvar data assimilation systems. Mon Weather Rev 125(9):2274–2292

Index

© Springer International Publishing Switzerland 2017
S. Lakshmivarahan et al., *Forecast Error Correction using Dynamic Data
Assimilation*, Springer Atmospheric Sciences, DOI 10.1007/978-3-319-39997-3

Printed in the United States
By Bookmasters